Resilient Health Care

VOLUME 3
Reconciling Work-as-Imagined and Work-as-Done

Resilient Health Care

VOLUME 3
Reconciling Work-as-Imagined and Work-as-Done

EDITED BY

Jeffrey Braithwaite
Robert L. Wears
Erik Hollnagel

CRC Press
Taylor & Francis Group
Boca Raton London New York

CRC Press is an imprint of the
Taylor & Francis Group, an **informa** business

CRC Press
Taylor & Francis Group
6000 Broken Sound Parkway NW, Suite 300
Boca Raton, FL 33487-2742

CRC Press is an imprint of Taylor & Francis Group, an Informa business

No claim to original U.S. Government works

Printed on acid-free paper
Version Date: 20160628

International Standard Book Number-13: 978-1-4987-8056-8 (Hardback)

Visit the Taylor & Francis Web site at
http://www.taylorandfrancis.com

and the CRC Press Web site at
http://www.crcpress.com

Printed and bound in the United States of America by Publishers Graphics,
LLC on sustainably sourced paper.

Contents

Part I Problems and Issues
Jeffrey Braithwaite, Robert L. Wears and Erik Hollnagel

Part II Applications
Jeffrey Braithwaite, Robert L. Wears and Erik Hollnagel

Part III   Methods and Solutions
Jeffrey Braithwaite, Robert L. Wears and Erik Hollnagel

Preface

Jeffrey Braithwaite, Robert L. Wears and Erik Hollnagel

Resilient health care (RHC) focuses on everyday clinical work (ECW) – which means work as it unfolds in practice – accomplished by those working at the sharp end of the system in direct contact with patients. But ECW takes place in conditions that are significantly shaped by those at the blunt end of the system, distanced from the sharp end in both time and space. Their aim is to ensure the safety and productivity of ECW by the way they prepare, organise and manage the conditions in which work takes place. Because today's health care takes place in complex sociotechnical systems, ECW relies on workarounds, trade-offs and adjustments in order for everyday activities to succeed.

In today's health care environment, with its complicated relationships, technologies, services and practices, it is inevitable that work-as-done (WAD) differs from work-as-imagined (WAI). ECW will therefore always be different from that which is intended, planned and prescribed. Although the differences usually are not dramatic, they may nevertheless at times lead to problems. In order to ensure that health care is resilient, it is therefore necessary to continuously realign the two perspectives on work, rather than insist on one being right (usually WAI) and the other being wrong (usually WAD).

That is the precise point where this book hopes to make a contribution. Before we introduce the book, however, we should provide some context for both the series to which it belongs, and for the background that has brought us to our current position.

The publication of this volume completes a serendipitous trilogy of work in RHC. To understand how this happened, we need to recognise the links to ideas that existed long before the current thinking about resilience and resilience engineering (RE) began in the 1970s. These ideas, as Hollnagel suggests in his Prologue, may even be glimpsed in Plato's stories, through Socrates' narratives, of the allegory of the cave – in which there are shadow-like appearances of people and animals on the walls of the cave (for us, WAI), and realities beyond the cave (for us, WAD). We have applied the modern version of these ideas to our enduring interest: the contemporary health system, replete with its complex idiosyncrasies, ambiguities, conflicting goals, socio-professional divides, competing interests, multiplicity of treatments, care models, services, staff and subsystems and enabling and constraining mechanisms.

RE pointed out that in the past, systems and safety experts had been pre-occupied with a "find and fix" perspective, focusing on things going wrong at the expense of understanding the whole gamut of performance. To break free from that, RE advocates that it is more important to understand how things work than how they fail, in order to be able to improve performance and increase the number of acceptable outcomes.

In our series of RHC books, we also recognise a link between the three volumes. In the first volume, *Resilient Health Care* (Hollnagel et al., 2013), we set the scene, and teased out the kinds of circumstances in which people on the front lines of care adjust, flex and accommodate in executing their functions as doctors, nurses and allied health professionals, and we glimpsed the relative contributions of both this sharp end and the blunt end to sustaining performance. Taking a lead from work that culminated in Hollnagel (2014b) the idea of two kinds of safety (Safety-I, aiming to reduce harm by trying to make sure things do not go wrong) and its reciprocal (Safety-II, aiming to improve the ability to succeed, and striving to make things go right) was teased out. The second volume, *The Resilience of Everyday Clinical Work* (Wears et al., 2015), concentrated much more directly on the front-line activities, and analysed many different settings – wards, departments, emergency care, critical care, operating theatres and community pharmacies, for instance – to reveal much more than was previously understood about the activities of those making such adaptations over time.

To complete the trilogy, we need to appreciate further the relationships between those who fund the services and specify the way care should unfold, and those who deliver the services in real time. In what follows, after this Preface from the three editors and a Prologue from Hollnagel, we begin the task of deepening our understanding of these relationships and arrangements.

To do this we have divided the chapter contributions into three sections. The first, *Problems and Issues*, covers topics as wide ranging as what resilience is and how people can absorb the ideas it has spawned, to the patient's role in creating resilient health care (and perhaps acting in a bridging role between WAI and WAD), to Lean approaches to improvement contrasted with resilient approaches. These initial chapters establish a platform for the work that follows.

The second section, *Applications*, shifts the emphasis. We present here an intriguing array of contributions articulating the ways we might bring together, or at least refine our appreciation of, WAI and WAD. The chapters touch on topics including WAD and health information technology, and the alignment of WAI and WAD in operating theatres. Other applications include how to model and represent WAD. Specific applications include manifestations of resilience in emergency care, and how systems that learn from everyday activities can contribute to narrowing the WAI–WAD nexus. Contributions also reflect on the point that power is never far away from the WAI–WAD paradigm, and, closely related to this, recognise that policymakers, managers, clinicians, patients and researchers not only have differing perspectives on RHC, but make different demands on the system that expresses it.

Armed with the information from the first and second sections readers will in the third section, *Methods and Solutions*, learn what to do with this information. This section distinguishes itself by advancing work contributing to bridging the gap between the WAI and WAD worlds: essentially, by researching, modelling or simulating activities; or by purposefully designing greater levels of resilience into the care processes wherever possible in order to strengthen rather than perturb them; or to train blunt- and sharp-end people in the utilities and capacities of resilience thinking.

One thing remains clear. It is not possible to provide a complete description of everything that goes on in a complex workplace. In that respect, WAI inevitably falls short of the aims it has for streamlined, effective and efficient work to unfold in pre-specified ways. But neither is it possible simply to expect front-line operators to get on with the job without providing them with resources, supporting systems and guidelines for performing. Both WAI and WAD are therefore essential: we simply cannot have one without the other.

Our core message is that whatever our vantage point, we ought not cling stubbornly to only one view of the world and insist that it is the right one. Whether we identify mostly with a regulatory-policy, managerial, top-down, blunt-end perspective, or whether our world is steeped in a front-line, practice-centric, bottom-up, sharp-end perspective, having only one standpoint is both insufficient and dangerous. Care needs to be planned, managed and funded. It also needs to be rolled-out, executed and enacted. It is when plans meet reality that the two worlds meet, are able to be reconciled and care is resilient. So it is in the synergy between the two that our hopes for better, more responsive, resilient, safer care must be based. The tensions between WAI and WAD, between standardisation and contingency, between sharp-end and blunt-end interests, cannot be resolved, but must be actively and creatively managed, case by case, through judicious actions by the parties involved.

Editors

Jeffrey Braithwaite, BA, MIR (Hons), MBA, DipLR, PhD, FAIM, FCHSM, FFPHRCP (UK), FAcSS (UK), is foundation director, Australian Institute of Health Innovation; director, Centre for Healthcare Resilience and Implementation Science; and professor of health systems research, Faculty of Medicine and Health Sciences, Macquarie University, Australia. His research examines the changing nature of health systems, attracting funding of more than AU$85 million (€54 million, £42 million). He has contributed over 600 total publications and presented at international and national conferences on more than 800 occasions, including 80 keynote addresses. His research appears in journals such as the *British Medical Journal*, *The Lancet*, *Social Science and Medicine*, *BMJ Quality and Safety* and the *International Journal of Quality in Health Care*. He has received numerous national and international awards for his teaching and research. Further details are available at his Wikipedia entry: http://en.wikipedia.org/wiki/Jeffrey_Braithwaite. He blogs at http://www.jeffreybraithwaite.com/new-blog/.

Robert L. Wears, MD, PhD, MS, is an emergency physician, professor of emergency medicine at the University of Florida and visiting professor in the Clinical Safety Research Unit at Imperial College London. His further training includes a master's degree in computer science, a 1-year research sabbatical focused on psychology and human factors in safety at the Imperial College, followed by a PhD in industrial safety from Mines ParisTech (Ecole Nationale Supérieure des Mines de Paris). He serves on the board of directors of the Emergency Medicine Patient Safety Foundation, and multiple editorial boards, including *Annals of Emergency Medicine, Human Factors and Ergonomics, Journal of Patient Safety* and *International Journal of Risk and Safety in Medicine*. Wears has co-edited three books, *Patient Safety in Emergency Medicine, Resilient Health Care* and *The Resilience of Everyday Clinical Work*, and he is working on two more. His research interests include technical work studies, resilience engineering and patient safety as a social movement. His research papers and commentaries have appeared in *JAMA, Annals of Emergency Medicine, Safety Science, BMJ Quality and Safety, Cognition Technology and Work, Applied Ergonomics* and *Reliability Engineering and Safety Science*.

Erik Hollnagel, MSc, PhD, is a professor at the Institute of Regional Health Research, University of Southern Denmark, chief consultant at the Centre for Quality, Region of Southern Denmark, visiting professor at the Centre for Healthcare Resilience and Implementation Science, Macquarie University, Australia, and professor emeritus at the Department of Computer Science, University of Linköping, Sweden. He has through his career worked at universities, research centres and industries in several countries and with problems from many domains including nuclear power generation, aerospace and aviation, software engineering, land-based traffic and health care. His professional interests include industrial safety, resilience engineering, patient safety, accident investigation and modelling large-scale sociotechnical systems. He has published widely and is the author or editor of 22 books, including five books on resilience engineering, as well as a large number of papers and book chapters. The latest titles, from Ashgate, are *Safety-I and Safety-II: The Past and Future of Safety Management*, *Resilient Health Care*, *The Resilience of Everyday Clinical Work*, *FRAM – The Functional Resonance Analysis Method* and *Resilience Engineering in Practice: A Guidebook*. Hollnagel also coordinates the Resilient Health Care Net (http://www.resilienthealthcare. net) and the FRAMily (http://www.functionalresonance.com).

Contributors

Janet E. Anderson
Centre for Applied Resilience (CARe)
Florence Nightingale Faculty of
    Nursing and Midwifery
King's College London
London, United Kingdom

Pierre Bérastégui
Department of Cognitive Ergonomics
University of Liège
Liège, Belgium

Jeffrey Braithwaite
Australian Institute of Health
    Innovation
Faculty of Medicine and Health
    Sciences
Macquarie University
Sydney, Australia

Carolyn Canfield
Department of Medicine
University of British Columbia
Vancouver, British Columbia,
    Canada

Karen Cardiff
School of Population and Public
    Health
University of British Columbia
Vancouver, British Columbia, Canada

Sheuwen Chuang
School of Health Care
    Administration
Health Policy and Care Research
    Center
Taipei Medical University
Taipei, Taiwan

Robyn Clay-Williams
Australian Institute of Health
    Innovation
Faculty of Medicine and Health
    Sciences
Macquarie University
Sydney, Australia

Lacey Colligan
Sharp End Advisory, LLC
Hanover, New Hampshire

Richard I. Cook
Integrated Systems Engineering
    Department
Ohio State University
Columbus, Ohio

Ellen S. Deutsch
Pennsylvania Patient Safety
    Authority
Harrisburg, Pennsylvania

Mirjam Ekstedt
Royal Institute of Technology
Medical Management Centre,
    LIME
Karolinska Institute
Stockholm, Sweden

Erik Hollnagel
Institute of Regional Health
    Research
University of Southern Denmark

and

Centre for Quality, Region of
    Southern Denmark
Copenhagen, Denmark

Garth S. Hunte
Emergency Medicine
University of British Columbia

and

Providence Health Care
Vancouver, British Columbia, Canada

Lisa Jacobson
College of Medicine
Center for Simulation Education and
  Safety Research (CSESaR)
University of Florida
Jacksonville, Florida

Peter Jaye
Emergency Medicine
Guy's and St. Thomas's NHS
  Foundation Trust
King's Health Partners Academic
  Health Sciences Centre
London, United Kingdom

Andrew Johnson
Townsville Hospital and Health
  Service
Queensland, Australia

Paul Lane
Townsville Hospital and Health
  Service
Queensland, Australia

Shinichi Masuda
Department of Clinical Quality
  Management
Osaka University Hospital
Osaka, Japan

Kazue Nakajima
Department of Clinical Quality
  Management
Osaka University Hospital
Osaka, Japan

Shin Nakajima
Osaka Graduate School of Medicine
Osaka National Hospital
Osaka, Japan

Anne-Sophie Nyssen
Cognitive Ergonomics and Work
  Psychology
University of Liege
Liege, Belgium

Mary Patterson
Children's Academy of Pediatric
  Educators
Children's National Medical Center
Washington, DC

Simone Pozzi
Deep Blue Consulting & Research
Rome, Italy

Alastair J. Ross
Behavioural Science
Glasgow Dental School
University of Glasgow
Glasgow, Scotland

and

Centre for Applied Resilience in
  Healthcare
King's College London
London, United Kingdom

Caroline Brum Rosso
Industrial Engineering Post-
  Graduation Program
Federal University of Rio Grande do
  Sul (UFRGS)
Porto Alegre, Brazil

Tarcisio Abreu Saurin
Industrial Engineering Department
Federal University of Rio Grande do
  Sul (UFRGS)
Porto Alegre, Brazil

Sam Sheps
School of Population and Public
  Health
University of British Columbia
Vancouver, British Columbia, Canada

Mark A. Sujan
Patient Safety
Warwick Medical School
Coventry, United Kingdom

Carlo Valbonesi
Deep Blue Consulting & Research
Rome, Italy

Robert L. Wears
Emergency Medicine
University of Florida
Jacksonville, Florida

and

Clinical Safety Research Unit
Imperial College London
London, United Kingdom

Prologue: Why Do Our Expectations of How Work Should Be Done Never Correspond Exactly to How Work Is Done?

Erik Hollnagel

Introduction

Discussions and arguments about work-as-imagined (WAI) relative to work-as-done (WAD) often take place in a polemical if not outright adversarial atmosphere. The purpose of the two terms is, of course, to highlight a distinction that is important for how work is managed – and *a fortiore* also for how safety, quality and productivity are managed – but in practice the terms are often used to assign responsibility or even blame for what many see as the lamentable state of health care in the industrialised societies today (Hollnagel, 2015a). Connotations and innuendos aside, the two terms do draw attention to an important but sometimes neglected problem, as many of the chapters in this and the previous volume demonstrate (Wears et al., 2015).

The current use of the WAI–WAD terminology started during and after the discussions at the resilience engineering symposium in Söderköping, Sweden, in 2004, as documented by Dekker (2006), with the public debate beginning soon after. But the realisation that there is a difference between the expectations to work and actual work practices is considerably older. One early source is the juxtaposition between work-as-prescribed (called the 'task') and work-as-performed (called the 'activity') that arose in French ergonomics in the mid-1950s, as described in Chapter 4, this volume. But the realisation that there is a difference between the world we can perceive – or imagine – and the real world is much older than that. It is possibly intellectually arrogant – or heretical? – to trace the origin back to Plato's allegory of the cave. It is true that WAI in some sense is a projection of WAD, like the shadows projected on the wall by things passing in front of a "fire behind them", but the projection is active rather than passive. It is what we want to see or expect to see, rather than what we would see if we, in Plato's terms, could perceive "the true form of reality".

We are on firmer ground if we jump forward to the beginning of the twentieth century. Leaving aside the rise of Scientific Management or Taylorism, one of the seminal examples is the so-called Hawthorne effect (Homans, 1958). From the late 1920s through the early 1930s the Western Electric Hawthorne works outside Chicago had commissioned a study to see if their workers would become more productive in higher or lower levels of light. Here the 'work-as-prescribed' or work-as-imagined was the 'obvious' hypothesis that there would be a direct relation between level of lights and productivity. Specifically, that improved illumination would lead to increased productivity – at least until a plateau was reached. But the expected dependency was not found. Further studies looked at the effects of other factors such as rest periods, mid-morning lunches and shorter working hours but were unable to find any simple relations. In order to explain the varied results, it was suggested that the (unsystematic) productivity gains occurred as a result of the motivational effect on the workers of the interest being shown in them. (In other words, the usual kind of explanation involving a single factor, which furthermore was created in hindsight.) Or as Mayo (1931) put it: "Somehow or other that complex of mutually dependent factors, the human organism, shifted its equilibrium and unintentionally defeated the purpose of the experiment" (p. 56).

The focus of the Hawthorne experiments was productivity rather than safety, but the lesson learned is still important. Namely that the researchers' (and presumably also the managers') understanding of what affected productivity did not correspond to what actually was the case. Reality was richer and more varied than their imagination. The parallel to safety is clear – that the idea of a specific intervention with a specific consequence will fail miserably if the understanding of what goes on, of WAD, is incomplete or incorrect.

Plans, Control and Management

The correspondence – or lack of correspondence – between WAI and WAD is important because work always must be managed, whether it is one's own or the work of others. In order to manage something it is necessary to understand what it is, so that one can select and apply the means that will bring the goal closer. More formally, control and management require three conditions to be met: first, it is necessary to have an adequate representation or model of how the local world works; second, it is necessary to know what the current status or conditions are; and third, it is necessary to know how desired changes can be brought about – and as part of that, when the results can be expected to arise. This is the case for any kind of managed activity, from driving a car in traffic to managing a hospital.

Plans provide the structure of behaviour (Miller et al., 1960). To prevent behaviour or performance from becoming random and ineffective, it is necessary to plan what is to be done in as much detail as possible. Indeed, everything we do is usually guided by plans, either implicitly (routine, habits) or explicitly (instructions, procedures). If there is no plan, if we have no idea about what we should do, then we say that control has been lost – as in the scrambled and opportunistic control modes (Hollnagel, 1998). To illustrate the ubiquitousness of plans, consider the following four situations:

- When I begin my working day, I know what I would like to accomplish during the day and I have a rough plan for how to do it. This plan, this exemplar of WAI, is, of course, a variation of what I usually do, depending on, e.g., the time of year, the deadlines, the backlog of work, etc. As the day progresses, the plan is revised several times, and the goals may also be revised. It is, indeed, unusual that I manage to accomplish everything I had intended to do. At the end of the day, WAD therefore rarely corresponds to what I had imagined from the beginning. What has been left undone is rescheduled, perhaps to the day after or perhaps to later – in the sanguine hope that the necessary resources will magically appear.

- When I drive, for instance, from the office to a meeting in another part of town, I also plan the travel with regard to, for instance, route and time. But often, my travel-as-done does not quite match my travel-as-imagined, simply because there are so many uncontrollable influences on this activity.

- In a different domain, Amalberti and Deblon (1992) over a 4-year period studied fighter pilots during high-speed, low-altitude penetration missions. One outcome of this study was the importance of preparing and planning for a mission, where two essential parts were the division of the plotted route into branches ('legs') of navigation to determine flight and navigation parameters, and a re-examination of each leg for possible threats. In this situation, the flight plan including the alternatives for possible threats represented the "mission-as-imagined", but with the clear realisation that the actual flight ("mission-as-done") usually would be different.

- Complex industrial services, such as power stations, data networks or train systems must have an availability as close to 100% as possible. Since it is important that their down-time is minimised, maintenance work is planned in minute detail. In a study of the procedures for testing before restarting a nuclear power plant after a period of maintenance, Gauthereau and Hollnagel (2005) identified three quite different types of organisational behaviour in as many weeks. In the first week work was performed in an orderly and methodical manner based on the planning of work before the outage period, although it was generally acknowledged that this covered only approximately one third of the jobs. In the second week the employees focused their attention on planning and on preparing the upcoming events in more detail, including responses to the 'surprises' found during the first week. Here people fell back on routines that had been learned (and mastered?) during previous outage periods. The third week was about 'following the plan'; more concretely, it was about adapting it to the circumstances so that the outage period could be ended on time. In this case it was clear to everyone that the prepared plans, WAI, would have to be adjusted more or less continuously to cope with the unexpected.

Although the four examples above are very different, they all represent a situation where WAI and WAD are simultaneous or contiguous, even if different people have different responsibilities. We can call this egocentric (*idiocentric*) or synchronous planning. It is egocentric because it is about the work that people do themselves, either as individuals or as teams. And it is synchronous because the plans are about what happens within the boundaries of the local work context and within the time horizon of the activity under consideration. This means that WAI and WAD are mutually coupled, hence that the plans can be corrected or adjusted based on the feedback from work.

Work-as-Imagined and World-as-Imagined

When we plan, we have to make certain assumptions about the world: how regular it is and how reasonable the people around us are. In other words how predictable, if not outright rational, the work context is. The more regular the world is and the more reasonable people are (leaving aside a discussion of the meaning of 'reasonable'), the more predictable the future will be and the easier therefore it is to plan what to do. Conversely, the more irregular the world is and the less reasonable people are the more difficult it is to plan what to do. When we plan, we try to anticipate what can happen almost as in a game of chess. But since few, if any, real situations are as constrained as a board game, and since the 'opponents' rarely behave as we imagined or even follow the rules, the plans will never correspond precisely to the situation.

Unless we have complete mastery and control of the world so that there are no differences between the world-as-imagined and the world-as-is, plans have to be adjusted to the conditions (or conditions have to be adjusted to the plan). The consequences of failing to do so were dramatically demonstrated during the fateful days of August 1914 when both the German and French armies followed plans (*Aufmarsch I* and *Plan XVII*) that had been worked out long before the war began (Tuchman, 1976). In the actual events, when the reality of the battlefield did not match the plans, the plans prevailed. As Tuchman dryly observed: "The Germans had worked out that the logical place for the

British to land would be at the ports nearest to the front in Belgium, and von Kluck's cavalry reconnaissance, with that marvellous human capacity to see what you expect to see even if it is not there, duly reported the British to be disembarking at Ostend, Calais, and Dunkirk on August 13 ... In fact, of course, they were not there at all ..." (Tuchman, op. cit, p. 254).

The problem is thus not just that WAI and WAD may differ, but also that what we think will be the conditions for work, or in this case a battle, the world-as-imagined, may be incorrect and quite unlike the world-as-is. If the assumptions about the conditions for work are incorrect, incomplete or over-simplified, then it is hardly surprising that the prescriptions or instructions for what to do also will be wrong.

The Law of Requisite Variety

The relation between the understanding of the situation and therefore the likelihood that plans will succeed and control can be maintained, has been formally addressed by cybernetics in the form of the Law of Requisite Variety (Ashby, 1956). This 'law' is concerned with the problem of regulation or control and expresses the principle that the variety of a controller must match the variety of the system to be controlled. In the current context this means that the contents of WAI should match WAD in terms of what the work conditions can be like, what can actually happen internally and exter-nally, and what the proper responses should be. We can think of the variety as all the different conditions that can possibly exist, or all the situations that may possibly occur. It stands to reason that in order to control a system, in order to be able to carry out the proper corrective or compensating actions, the controller must have at least the same variety or richness as the system to be controlled. Or to put it more simply, if something happens in practice that has not been considered when the practice was planned and the responses prepared, then performance will suffer.

According to the Law of Requisite Variety, effective control is only possible if the controller has more variety than the system. A situation can therefore be brought under control either by increasing the variety of the controller or by reducing the variety of the process (the situation at work). The former can be achieved by providing people with the necessary skills and competence, for instance through training or simulation (cf., Chapter 13). The latter is the 'logic' behind attempts of standardisation and an insistence on compliance. Standardisation can reduce the variety of the processes (as in quality assur-ance) and emphasis on compliance and following instructions can reduce the variety of the people in the system. This, however, misses the point that managing the people at the 'clinical coalface' is not the only concern. The clinical staff must also themselves manage something, for instance the patient's health. And since health problems – and patients – are notably dif-ficult to standardise, the solution of forcing or constraining WAD to match WAI is no solution at all.

Corresponding to requisite variety, Westrum (1993) has proposed the notion of *requisite imagination*. The context was accidents in complex systems, and requisite imagination was described as the ability to speculate about the possible ways in which something can go wrong. But it is natural to extend the use of the term to the planning of work in general. Requisite imagination applies to the way we think about work that is yet to be done, to the future, but both to the work where we ourselves are engaged and to work that others do and in which we do not ourselves take part.

Allocentric WAI–WAD

As discussed above, any differences that may exist in egocentric WAI–WAD can in principle be reconciled because WAI and WAD are contiguous in space and time. The consequences of any differences can therefore quickly be noticed, and can be fed back to the plans that structure behaviour.

But there are other situations where WAI and WAD are separated in space and in time. This can be called allocentric or asynchronous WAI–WAD. It is allocentric because it is not about the work that people do themselves but about work that others do; the plans are therefore made away from the actual place of work. And it is asynchronous because the plans are made before – and sometimes long before – work is being carried out.

Asynchronous WAI–WAD corresponds to the differences between the sharp end and the blunt end, as described in Hollnagel (2015a). The problem is, however, not just the polemic clash between the sharp and the blunt ends, but rather the impossibility of predicting or expecting how work that is done by others at a different time and in a different place will unfold in practice. In such cases there are no real possibilities for feedback and correction, and therefore few opportunities for learning. People at the (relative) blunt end do try to imagine or understand what WAD – and the world-as-is – will be like. But there are many things that make this difficult, primarily lack of time and lack of information. Because the world is a 'blooming, buzzing confusion' (James, 1890) made up of countless, interconnected systems, we resort to approximate adjustments in our reasoning. This has been characterised in several ways such as Merton's (1936) discussion of the "Unanticipated Consequences of Purposive Social Action", Simon's (1956) description of *satisficing*, and Lindblom's (1959, 1979) characterisation of 'muddling through'.

Incrementalism or 'Muddling Through'

The essence of 'muddling through' – as well as of *satisficing* – is that people do not make decisions by carefully identifying alternatives, comparing alternatives and finally selecting the optimum one. Because most actual situations have incomplete, dynamically changing conditions and competing goal structures, people tend to make their decisions by defining the principal objectives, outlining the few alternatives that occur to them, and finally selecting the one that seems to be a reasonable compromise between means and values. Although this has been studied mostly for the decisions that are part of work at the sharp end, for WAD, it obviously also applies to the decisions that are made at the blunt end. In other words, when people think about work and particularly when they are planning the work of others, they also tend to 'muddle through'.

Lindblom argued that 'muddling through', which also goes by the more formal name of incrementalism, should not be seen as the failure of a rational method but rather as a method or a system on its own. His studies showed that it was "... in fact a common method of policy formulation, and (is), for complex problems, the principal reliance of administrators as well as of other policy analysts" (Lindblom, 1959, p. 88).

In his later analysis, Lindblom (1979) described several important features of incrementalism. One is that the analysis is limited to a few somewhat familiar policy alternatives. The imagination is in other words constrained by that which is known and familiar. Another is that there is "an intertwining of analysis of policy goals and other values with the empirical aspects of the problem" and that the analysis only explores some but not all of the possible consequences of a considered alternative. This is more or less the same as Merton's (1936) description of the imperious immediacy of interest, by which he meant that the decision maker's "paramount concern with the foreseen immediate consequences excludes the consideration of further or other consequences of the same act". The decisions are therefore biased and may even come close to being wishful thinking. A third is that there is more focus on the "ills to be remedied than positive goals to be sought" (Lindblom, 1979, p. 517) – or in other words a preoccupation with Safety-I issues. And finally, there is a fragmentation of analytical work to many (partisan) participants in policy making, which makes it more likely that the overview is lost than maintained.

What to Do?

The possible solutions to reconcile the differences between WAI and WAD are not the same for egocentric and allocentric WAI–WAD. In the case of egocentric WAI–WAD, the focus of work planning, of WAI, is "what I will do" or "what we will do". The work in question is therefore done by the same people or the same group of people who plan the work or think about the work. This means that there it is possible to have a direct coupling, a working feedback, from the work that is done to the plans for work, and therefore also to adjust to those plans on a regular basis. The difference can therefore in principle be overcome or neutralised because egocentric WAI–WAD is synchronous.

In the case of allocentric WAI–WAD, the focus of work planning, of WAI, is "what they will do" or "what someone will do". This means that there is a 'distance' between the planning and the actual work in both time and space. This 'distance' can be considerable, to the extent that the people at the sharp end have no idea about who the planners were (or are) or about when – and how – the plans were made. The extreme case is obviously when work is governed by rules set down by legislators and/or regulators. In such cases the time lag may easily be years. When plans (regulations, guidelines) eventually are revised, it may be a whole new group of people who are involved. Under such conditions it is very hard to reconcile WAI and WAD in a meaningful sense. Allocentric WAI–WAD is a serious problem, which only gets worse as the complexity of complex sociotechnical systems (CSS) continues to grow.

The bottom line is that the expectations or descriptions of how work should be done never will match how work actually is done. Not necessarily because of ill will, arrogance or ignorance, but simply because there never will be enough time and enough information. We cannot rationally analyse work and prescribe the optimal configuration of ways and means except by constraining the descriptions to such an extent that they become completely unrealistic. Yet we do need to reconcile the potential and actual differences between WAI and WAD. Without that it is impossible to manage work so that quality, productivity, safety, etc., achieve an acceptable level.

Some ideas about how to reconcile allocentric WAI–WAD may be found in Westrum's (1993) suggestions for how requisite imagination might be improved. One way would be by providing incentives for thought and by having the organisation encourage individuals to think and be creative, in other words, to adjust the efficiency-thoroughness trade-off towards greater thoroughness. A second way was to remain open to ideas and to battle the "not invented here" syndrome. Finally, information flow should be facilitated by supporting openness to information and fostering a non-punitive attitude – which we today call a "just culture".

A comparable set of suggestions can be found in Righi and Saurin's (2015) proposed guidelines for how to manage complex sociotechnical systems. One suggestion is to improve the visibility of processes and outcomes by paying more attention to the informal work practices that over time become part of normal work. A second suggestion, which resembles Lindblom's reflections on incrementalism, is to encourage diversity of perspectives when making decisions. Diversity of perspectives may prevent some of the unwanted consequences of Merton's 'imperious immediacy of interests'. A third suggestion is to monitor and understand the gap between prescription and practice, particularly for allocentric WAI–WAD cases. Instead of simply assuming that changes will have the desired effects – unless people 'defeat' the good intentions – the consequences of small changes should be closely anticipated and monitored. As small changes happen all the time, they offer frequent opportunities for reflection on practice.

The difference between WAI and WAD may well be unavoidable, but it is not unmanageable. It can, however, only be managed if we recognise its existence and understand the reasons for it. The single most important reason is the human tendency to trade off thoroughness for efficiency. This is the reason why solutions often are incompletely thought through, and why we accept oversimplified descriptions as the basis for our plans and analyses. But we do so at our peril.

Part I

Problems and Issues

**Jeffrey Braithwaite, Robert L. Wears
and Erik Hollnagel**

Can an image of something, or a representation, or a plan of it, correspond to the thing itself? Alfred Korzybski once argued, no: the map is not the territory (Korzybski, 1931). The great philosopher Immanuel Kant also thought not. Although his theory of the world is much deeper and more extensive than we need here, for our purposes he made the claim that there are noumena (or abstract things that subsist but which we can never fully know) and phenomena (or the reality of things, displaying themselves to the perceiver through his or her senses). Essentially, Kant saw objects in two modes – things-in-themselves and things-as-they-appear (Kant, 1781, transl. Guyer and Wood, 1998). This vexed and vexatious idea has challenged philosophers before Kant. As Hollnagel points out in the Prologue, Plato's Theory of Forms postulated 2500 years ago a distinction between idealized abstractions that are in the mind, and direct knowledge known through the senses. In our domain of resilient health care, work-as-done (WAD) represents the observable phenomenon, or things-in-themselves, and work-as-imagined (WAI) represents the abstract noumena, or things-as-they-appear.

Although Korzybski, Plato and Kant had a different purpose than us (to do philosophy, compared with our more practical aims, to conceptualize ways to improve performance and safety), by way of analogy, this is the distinction that we are examining here. Our distinction is between the 'imagined' world of politicians, policy makers, managers, researchers and software designers and the 'concrete' world of doctors, nurses and allied health professionals, ward clerks and porters, each of whom experiences and contributes to care on those front lines. The former rely on their understanding or imagination of how work should or could be done to influence how work is done or can be done on the clinical front lines.

So the broad task we set ourselves, to reconcile views that imagine how work unfolds with those who do that work, has in different guises troubled intellectual giants for millennia. But before we can proceed very far in this excursion into the WAI–WAD spheres, we must begin to specify some of the problems and issues we encounter.

This brings us to the first five of our chapters, which collectively help us to secure a sound footing on which to base the rest of the book. We have grouped these under this first section, labelling it Problems and Issues. These chapters help tease out the scope of some of the challenges facing us, and the tests these set for the remainder of the book.

What types of problems and issues arise in these early chapters in clarifying the WAI-WAD distinction? A central one is whether we can, with purposeful effort, contribute to the task of actually conjoining the distinct WAI–WAD perspectives. That is firmly what the chapters in this section help us to think through. They range from Saurin, Rosso and Colligan (Chapter 1) and Sheps and Cardiff (Chapter 2), both of which help us to appreciate distinctions between attempts to create more Lean, streamlined and efficient health care in contrast to strengthening processes and resilient engineering perspectives in order to support naturally occurring flexibility and variation, to the role of patients in acting as a go-between across the WAI–WAD perspectives (Canfield, Chapter 3); to Nyssen and Bérastégui's Chapter 4, on the discrepancies and similarities between individual and systems resilience; and to the implementation of greater levels of resilience by harnessing Safety-I and Safety-II strategies, as presented by Chuang and Hollnagel in Chapter 5.

Altogether, these five chapters delve into these challenges in considerable detail. They provide some fundamentals for the chapters that follow. They do not identify every problem or issue at the nexus of WAI–WAD, of course. But they offer more than enough foundational work on which the later chapters build.

1

Towards a Resilient and Lean Health Care

Tarcisio Abreu Saurin, Caroline Brum Rosso and Lacey Colligan

CONTENTS

In this chapter, two approaches to managing health care systems are compared: resilience engineering (RE) and Lean production. Both approaches use modelling methods built on assumptions that influence the resulting analysis and any recommendations that are made. A comparison is made between representative methods of RE and Lean: the functional resonance analysis method (FRAM) and the value stream mapping (VSM) method, respectively. A study of the process of administering medications for patients hospitalized in an emergency department provides the empirical basis for this analysis. Reciprocal learning opportunities between the two methods are identified.

Introduction

Increasing economic, social and demographic pressures on health care organizations have fostered the adoption of process improvement methods found effective in other sectors. In particular, the use of Lean production, originally developed in auto manufacturing, has been embraced as an effective way to eliminate waste by concurrently minimizing supplier, customer and internal variability (Shah and Ward, 2007). The popularity of this approach has led to the evolution of "Lean health care", and Lean is now considered an important tool for health care improvement (Kenney, 2011; Spear, 2005). Concurrently, the interest in RE has grown in health care (Fairbanks et al., 2014; Hollnagel et al., 2013). In contrast with Lean, RE research has shown the value of performance variability by front-line practitioners. This variability is seen as essential for the safe care of patients and for contexts that are replete with uncertainty (Sujan et al., 2015c; Nyssen and Blavier, 2013). In a cursory view, the two approaches seem to be in conflict, as they place different emphasis on variability and have different goals: the goal for Lean is efficiency and the goal for RE is safety. However, earlier studies have identified theoretical synergies between Lean and safety management approaches based on complexity insights, such as RE. For example, "use visual controls", which is a core Lean principle (Liker, 2004), can be useful for resilience since it makes complexity visible and supports performance adjustment (Saurin et al., 2013). Moreover, both RE and Lean recognize there is a difference between work-as-imagined (WAI) and work-as-done (WAD). WAI is what designers, managers, regulators and authorities believe happens, or should happen. WAD is what actually happens in the workplace. Individuals and organizations are always adjusting to current conditions, constraints and the context of the moment. Thus, WAD reflects performance variability that underlies successful execution of "work in the wild" (Hollnagel, 2014a). Both RE and Lean recognize that the gap between WAI and WAD results from system complexity, rather than from faulty human performance (Hollnagel, 2014b; Liker, 2004).

In this chapter, a comparison is made between RE and Lean methods to describe sociotechnical systems (CSSs). The manner in which a system is described influences the analysis and recommendations made. The functional resonance analysis method (FRAM), which derives from the complexity and RE tradition (Hollnagel, 2012a), and the value stream mapping (VSM) method, which derives from Lean (Rother and Shook, 1998), are compared. FRAM has been the prominent tool for modelling complex CSSs in line with RE premises and provides insights as to how variability propagation affects performance. VSM identifies the extent to which the system is aligned with Lean principles and identifies processes that would benefit from Lean practices. A study of the process of administering medications in patients hospitalized in an emergency department (ED) provides the empirical basis for this comparison.

Research Design

A field study of medication administration was conducted in the ED of a major university hospital in the south of Brazil. Since the late 1970s, the ED has provided care in general practice, general surgery, gynaecology and paediatrics on a 24/7 basis. On average, 150 medical consultations are performed each day, totalling about 54,500 annual encounters. The ED has 275 employees distributed across 12 professional categories, and the physical environment is divided based on the level of patient acuity. The ED is part of the public health care network and is usually overcrowded. Although the official capacity is 41 adult and 9 paediatric beds, the number of patients is usually two or three times higher. Stretchers placed in corridors and hallways are commonplace, and management often uses local media to announce that the ED is closed due to overcrowding. The shortage of inpatient beds in the hospital is one of the contributing factors to overcrowding, and it is common that patients are cared for in the ED, frequently over several days.

The process of administering medications was selected for the FRAM and VSM comparison, because it is central to both patient safety, a core RE concern, and efficient supply of equipment and medications, a core Lean concern. This ED had previously been studied by the same research team and the work included a characterization of existing complexity attributes and the assessment of guidelines for managing complexity (Righi and Saurin, 2015). The VSM was applied first because the researchers were relatively more familiar with that method, rather than due to any identified relation of precedence. The researchers used a database developed in the previous work, which involved interviews with 18 members of staff, about 110 hours of observations of work at the frontline over a period of 6 months and the analysis of a number of documents, such as standardized operating procedures and reports of adverse events. This database supported identification of the ED functions, informal working practices and sources of variability.

The VSM Model

VSM was applied based on the guidelines proposed by Rother and Shook (1998), which involve four steps: (1) define a family of products or services to be mapped, (2) design the map of the current state, (3) design the map of the future/desired state and (4) develop an action plan to implement the future state. In this study, only (1) and (2) were addressed, as the further steps (3) and (4) are still pending agreement by ED top management. Two high-use medications (heparin and dipyrone) were selected as the family of products to be mapped (step 1), which means that for the purposes of the VSM these two medications could be regarded as a whole. A family is defined by shared characteristics of products or services, such as similar processing stages and processing times.

The map of the current state (step 2) was developed based on observations and informal conversations with employees. This map (Figure 1.1) has three broad zones. On the top, the information flows that support the

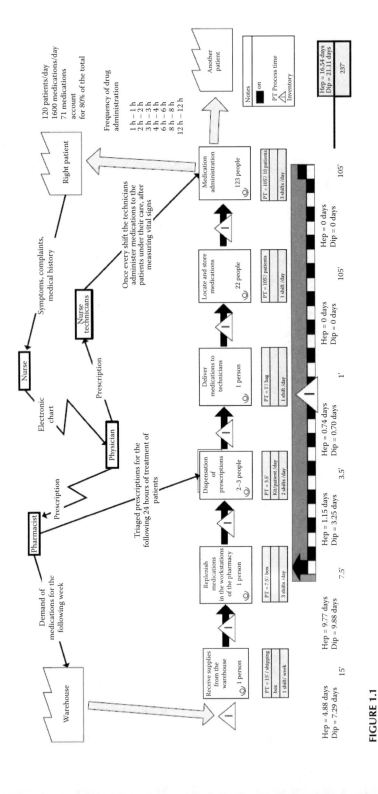

FIGURE 1.1
Map of the current state: flow of administration of medications.

physical flow of medications are represented. The boxes with borders in bold, on the top, refer to the agents exchanging information, and the nature of the information is between the arrows. Lines in zigzag mean electronic information flow, while straight lines mean physical flow of information. The icon on the top left represents the upstream boundary of the analysed system, which in the present case is the main warehouse of the hospital, from which the medications are transported to the pharmacy within the ED once a week. The icon on the top right represents the downstream boundary, which is the final destination of the medications. According to Figure 1.1, the medication can be administered either to the right patient, or if he or she is not found/has been discharged, to another patient who needs the same medication.

The middle portion of the map represents the physical flow of medications. The stages of the physical and information flow have supplier–client relationships, and the client requirements are identified. For instance, the pharmacy is regarded as a client of both the physicians who prepared prescriptions and the central warehouse of the hospital, which supplied medications and equipment. Because the nurse technicians picked up medications from the pharmacy, they are regarded as its clients. In turn, patients are regarded as the clients of technicians, who administered the medications. Data to describe each stage of the physical flow are collected, such as processing times, waiting times, the number of employees involved in each stage and their professional categories. The boxes correspond to the stages in which professionals are handling the medication, and the triangles represent stocks of medication. The horizontal arrows indicate the direction of the physical flow, and the zebra by convention means that this is a "push system". Push production is the opposite of the Lean ideal of pull production, or just-in-time (Liker, 2004), and implies there is no work-in-process cap for the stocks of medications, and that each stage of the stream works as a silo, doing its work and 'pushing' it to the next stage independently of its real needs.

The bottom part of the map is the timeline, which shows the lead time of one unit (or batch) of medication from the stock within the pharmacy at the ED to the patient. In the present map, the medications always are processed in batches throughout the value stream – e.g., they are delivered to technicians in a bag that contains all medications for the patients under their care. The technicians usually administer medications to batches of 10 patients. Timeline data allow comparison between processing time and total lead time. In this study, it was found that the total lead time of the medications was fairly long (about 21 days for dipyrone), although just 0.4% of this time (105 minutes) was spent on value-added activities – i.e., administering the medication. Similarly, it was observed that the value stream stage "locate and store medications" took about 105 minutes every day, consuming precious time from nurse technicians.

The FRAM

Execution of the FRAM followed steps from Hollnagel, Hounsgaard and Colligan (2014) and Hollnagel (2012a): (1) to define the purpose of the FRAM analysis; (2) to identify and describe the functions, which refer to the acts or activities that are needed to produce a certain result; (3) to identify variability; (4) to identify the aggregation of variability and (5) to identify consequences of the analysis. Researchers decided that the purpose of the FRAM (step 1) would be a risk assessment of administering medications in a 'normal' scenario of overcrowding. The database developed by Righi and Saurin (2015) and the map of the current state were adopted as a starting point for identifying and describing the functions (step 2). However, that database was not focused on the administration of medications, and some sources of variability were either absent or hidden in the information flow portion of the VSM. As a result, three interviews with pharmacists and nurse technicians, as well as 20 hours of observations of the process of administering medications, were conducted specifically for application of the FRAM. Table 1.1 presents the identification of the *potential* variability (step 3), involving an

TABLE 1.1

Excerpt from the Identification of Variability

Function	Output	Variability of the Output
Prescribe treatment	Prescription is provided for the following 24 hours of treatment; exams are requested.	Too late: an agreement between the pharmacists and the physicians establishes that all prescriptions should be delivered no later than 4:00 p.m., but prescriptions are usually delivered later than that. Acceptable: prescriptions are usually correct, although some information may be missing.
Check prescription	Prescription checked: several aspects of prescriptions are checked, such as dose and the risks arising from the combination of medications.	Too late: as the delivery of prescriptions to the pharmacy is not evenly distributed throughout the day, peaks of workload may delay the checking of prescriptions. Imprecise: the workload peak increases the chances of missing errors in the prescriptions.
Dispense medications	Medications dispensed for the next 24 hours of treatment for each patient: medications for each patient are placed in a small bag, identified with the name of the patient and their location.	Too late: delays in the upstream functions may impact on this function. Imprecise: it is common that the patient is not in the treatment unit assumed by the physician when he or she made the prescription; thus the dispensed medications will be placed in the wrong bag. Also, changes in the clinical condition of the patient over 24 hours may render the medication no longer necessary, or the prescription may be outdated.

(Continued)

TABLE 1.1 (*Continued*)

Excerpt from the Identification of Variability

Function	Output	Variability of the Output
Deliver medications to technicians	Dispensed medications are delivered to nurse technicians – a larger bag contains the smaller bags with the medications of all patients from each treatment unit within the ED.	Too late: delays of the upstream functions may impact on this function. Imprecise: as mentioned above, the bags received by technicians are unlikely to contain all the medications for the patients under their care. In fact, the bag usually contains medications of patients who are in other units within the ED.
Locate and store medications	Medications located and stored in each treatment unit: each nurse technician has to take all medications out of the larger bag, and find the patients under their care.	Too late: in addition to the delays in the upstream functions, the manual work of taking all medications out of the bag in order to find the right ones is a time-consuming activity. Imprecise: the number of drawers to store the medications is lower than that of patients. Consequently, medications may be stored in improvised locations without clear identification.
Locate patients	Patients located: before administering medications, technicians need to locate the patients.	Too late: in addition to the delays in the upstream functions, nurse technicians spend time looking for the patients under their care. The dynamics and overcrowding of the ED affect performance of this function. Imprecise: there were reports of misidentification of patients with the same name, which led to near misses.
Measure vital signs	Vital signs are measured: usually, blood pressure, temperature, and pain.	Too late or not at all: due to overcrowding and time pressure created by delays in upstream functions, nurse technicians sometimes only measure vital signs of the more critical patients. Delays also happen because there are insufficient numbers of equipment available. Precise: most vital signs are measured with the support of reliable and specific equipment.
Administer medications	Medications are administered.	Too late or not all: it is fairly common that medications are administered for some patients a few hours later than the prescribed time. In rare cases, administration may be missed completely. Imprecise: because monitoring is missed for some patients, there is a risk of administering medications that should be withheld due to worrisome vital signs – e.g., a medication for lowering blood pressure may be administered to a patient who has a low blood pressure at the time of administration.

TABLE 1.2

Excerpt from the Aggregation of Variability for the Risk Assessment

The variability of the output of the function <Prescribe treatment>	
May propagate to the function <Check prescription>	
Affecting one or more of the aspects below – explain when and how	
Input (I)	Prescription made is an input for <Check prescription> in the pharmacy. If the prescription is made too late, this increases the variability of the input of the downstream function.
Time (T)	No identified coupling
Precondition (P)	No identified coupling
Control (C)	No identified coupling
Resource (R)	No identified coupling

Source: Adapted from Von Buren, H., Unpublished MSc Thesis, Federal University of Rio de Janeiro, Rio de Janeiro, Brazil, 2013.

analysis of how the outputs of each function could vary in terms of time and precision, from the perspective of the needs of downstream functions. For instance, a precise output satisfies the needs of the downstream function, whereas an imprecise output is incomplete, inaccurate, ambiguous or in other ways misleading (Hollnagel, 2012a). The analysis of potential variability was not restricted to any specific scenario. Table 1.2 illustrates how the aggregation of variability (step 4) was made. In this step, researchers had to assess whether the *actual* variability of the output of any given function, considering the previously mentioned scenario of overcrowding, affected any aspect of any other function. Actual variability refers to what should realistically be expected to happen in a given scenario (Hollnagel et al., 2014). As recommended by Hollnagel (2012a), both everyday and unusual variability were accounted for steps (3) and (4). Whenever an output of one function provided an aspect of another function, a coupling between two functions was established. The couplings are graphically represented in Figure 1.2, which illustrates the instantiation of the FRAM for the analysed scenario, through the FRAM Model Visualizer 2.0.

Comparing FRAM and VSM

The results obtained in these particular analyses allow comparison of the FRAM and VSM model. The VSM offers useful attributes, such as (1) the notion of 'internal clients', their requirements and what counts as acceptable performance; (2) an emphasis on quantitative data to describe each stage of the value stream, as these data may be required to justify necessary changes to managers and (3) information about what happens between the stages of the value stream (these stages were assumed as equivalent to the FRAM functions) – e.g., medications waiting to be used by the downstream processes.

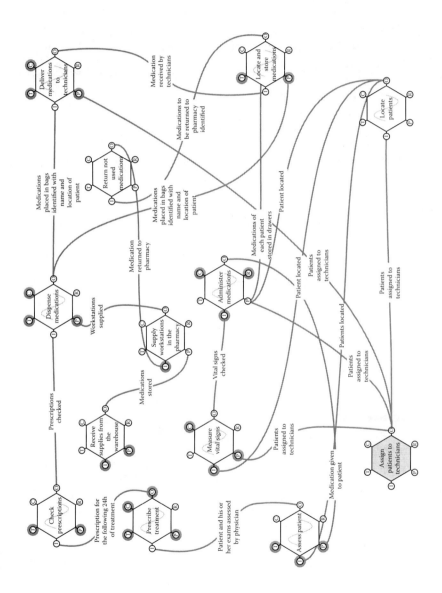

FIGURE 1.2
Instantiation of the administration of medications for a typical scenario of overcrowding.

Concerning FRAM, it provides insight into why WAI may not match WAD. This is mostly due to the qualitative emphasis of the FRAM, which accounts for the context and sources of variability that may affect requirements of clients. In fact, although Lean encourages managers to go to the *gemba* (in Lean jargon, "the real place, where the work is actually done"; Liker, 2004), VSM does not support sense-making of the gap between WAI and WAD. Literature oriented to Lean practitioners (e.g., Martin and Osterling, 2014) often conveys oversimplified messages, such as "walk the value stream" and "talk with operatives"; however, this may be insufficient for grasping WAD. In fact, some functions were not identified by VSM, such as <Locate patients>. This was because VSM focused on what the elements being mapped (i.e., the medications) were doing, rather than on what people were doing. The medications were waiting (i.e., they were stored either in drawers or somewhere else) while <Locate patients> was performed, so at this moment, the flow of people and the flow of materials were disconnected. That being said, it is worth noting that the flow of medications is just one of the flows that could have been mapped through VSM. If the flow of providers had been mapped, <Locate patients> would naturally appear in the VSM. As a drawback, although there are several relevant flows in health care settings (e.g., patients, physicians, medications), Lean literature does not offer guidance on how to select the flows to be mapped, or how to analyse their relationships (Black, 2008).

FRAM also identifies local variability and its system-wide impact. For example, the FRAM instantiation (see Figure 1.2) indicates how the timely and precise execution of the function <Administer medications> depends on timely and precise outputs from several upstream functions, which do not necessarily happen sequentially. Similarly, FRAM allows the identification of systemic impacts of re-designing individual functions. For example, if medications were dispensed in smaller batches (e.g., for the following 6 or 12 hours, rather than for the following 24 hours), the output of <Dispense medications> would be probably less variable in terms of time and precision, thus benefiting almost all other functions. Even apparently unrelated functions such as <Assess patient> would be impacted, because timely and precise medication administration improves the health condition of the patient, reducing the time until the patient can be discharged. The main point conveyed by this example is that the beneficial impacts of relatively small changes, such as working with smaller batches, may have significant effects throughout the system due to multiple couplings. This can explain why the Lean principle of doing many small and continuous improvements (i.e., kaizen) can produce significant results.

Table 1.3 summarises the comparison between VSM and FRAM, indicating that they manage the trade-off between ease of use and thoroughness in different ways. Overall, the FRAM provides a more nuanced description of the system, but data collection and analysis may be more arduous. However, as the field study presented in this chapter indicates, VSM portrays the system as more linear than it really is, thus conveying at least two misleading

TABLE 1.3

Comparison between FRAM and VSM

	FRAM	VSM
General Criteria		
Origin	Studies of safety and resilience in CSSs	Studies of Lean in manufacturing
Typical use of the method	Risk assessment and accident investigation	Mapping the current state and designing the desired future state. It sets a basis for the identification of which Lean practices may be useful and where they should be used
Performance dimensions emphasized	Safety and variability	Lead time and efficiency
Unit of analysis	Functions: the acts or activities that are needed to produce a certain result (Hollnagel et al., 2014)	Stages of the value stream: it includes both adding-value and non-adding-value stages
How is the unit of analysis described?	Six aspects of functions: input, output, preconditions, resources, time and controls	Several data points required (e.g., cycle time, set-up time, efficiency, number of workers), although there is no standardized set of data
Is visibility given to what exists in-between the units of analysis?	No, the method simply states that functions are connected to each other through variability propagation	Yes, the typical assumption is that there are queues/work-in-process in between the adding-value stages
Assumption on the nature of interactions between system's elements	Nonlinear	Linear
Does the method require the use of quantitative data?	No	Yes
Extent to which the method is underspecified in the seminal books and papers	Moderate: seminal publications (e.g., Hollnagel, 2012a) do not present a detailed prototypical case, encompassing all stages of the method	Moderate: seminal publications (e.g., Rother and Shook, 1998) focus on prototypical cases in linear process of the manufacturing industry
Degree of dissemination of the method among practitioners	Low, but increasing	High, especially in the manufacturing industry
Steps for application of the FRAM and VSM		
Scope	To define the purpose of the FRAM analysis	To define a family of products or services to be mapped
Model of WAD	To identify and describe the functions, to identify variability and to aggregate variability	To design the map of the current state

(Continued)

TABLE 1.3 (*Continued*)

Comparison between FRAM and VSM

	FRAM	VSM
Model of WAI	To identify the consequences of the analysis	To design the map of the future state and to develop an action plan to implement the future state

messages: (1) that variability in relation to the graphical representation is unimportant, which conflicts with the notion that CSSs quickly evolve over time, and that small changes should not be overlooked; and (2) that the mapped value stream has no relevant interactions with other elements of the CSS. This is in conflict with the fact that CSSs are open systems, formed by a large number of dynamically interacting elements (Cilliers, 1998).

What Can FRAM Practitioners Learn from the VSM?

The VSM provided some insights into the FRAM. First, when identifying functions, attention should be paid not only to what people, organizations and technologies *do* in order to produce a certain output. The time when people, organizations and technologies *do not do anything*, i.e., when they are waiting, should also be considered. This is an important issue in health care, as patients may often wait for a long time before being cared for.

Second, the explicit identification of client requirements by Lean and VSMs might provide focus for the last step of the FRAM (i.e., to identify consequences of the analysis) as well as support the analysis of the variability of the outputs, in regard to time and precision. For example, when doing the VSM, it was explicit that the patient was the client of the process of administering medication, and that his or her requirements could be summarised by the five rights – e.g., right time, medications should be administered at 8:00 p.m. every day. Such information defines outputs that are too early, on time or too late. Furthermore, identification of client requirements helps us to identify functions that do not add value (e.g., <Locate patients>, and <Locate and store medications>), which could be either eliminated or reduced through better system design. Third, the VSM's quantitative data, such as processing times and rate of consumption of resources by clients, provide data for describing the aspects of the functions, such as how often outputs are produced ('time' aspect). This may be useful for the identification of windows of opportunity for the propagation of variability and, as a result, when variability should be monitored. Quantitative data also address the demand and capacity for each function. A mismatch is a source of variability and may mean that essential slack is being consumed, and resilience is being reduced.

What Can VSM Practitioners Learn from the FRAM?

By analysing how the FRAM describes variability and WAD, VSM practitioners could recognize that going to the gemba is not the same as understanding it. In fact, RE and the FRAM might work, respectively, as a theoretical framework and an analytical tool for supporting VSM practitioners in making sense of what is going on at the gemba. Furthermore, by using the FRAM as a risk assessment tool, VSM practitioners can evaluate the broader impacts of typical Lean solutions that are designed in the maps of the future state. For instance, the FRAM indicates, as previously discussed, the consequences of dispensing medications in smaller batches. Similarly, the FRAM can support the analysis of the impacts of a hypothetical just-in-time (JIT) delivery of medications to the pharmacy. According to the FRAM instantiation, JIT delivery of medications could make the function <Dispense medications> more susceptible to variability in the output of <Prescribe treatment> – e.g., batches of prescriptions involving unusual amounts and types of similar medications (for instance when dealing with victims of disasters) would be unlikely to be matched by the lower stocks. Thus, it seems that the effectiveness of the solutions designed into the map of the future state may be inspected through a risk assessment using the FRAM.

Last, the analysis of variability carried out through the FRAM might be useful to the assessment of how stable the stages of the value stream are. In fact, some Lean practices, such as pull production/JIT, require stable processes along the value stream. Otherwise the upstream process will not be able to supply its client at the right time, at the right amount. Stability in Lean production is usually framed in terms of the four Ms – man, machine, methods and materials (Smalley, 2004).

Lean or Resilience?

This comparison requires a framework for the identification of whether Lean or resilience should be emphasised in the re-design of CSSs. The proposed framework (Figure 1.3) has two dimensions: the complexity of the system being analysed, and whether the attributes of complexity are either a liability or an asset for the system. Owing to the lack of widely accepted methods for measuring the complexity of CSS, this framework is intended to support negotiations between stakeholders, rather than to precisely define systems. Of course, estimates of complexity can be made, such as by identifying the extent to which widely agreed attributes associated with complexity (e.g., uncertainty, dynamism, diversity) exist in the system (Righi and Saurin, 2015). For situations in quadrant (I), resilience should be emphasised in order to cope

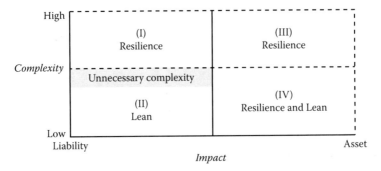

FIGURE 1.3
Framework for identifying whether Lean or resilience should be emphasized.

with high complexity. In quadrant (II), Lean should be emphasised in order to exploit the benefits of linearity. This quadrant also recognises that a portion of unnecessary complexity exists, due to badly designed processes that could be made linear through the use of Lean practices. Quadrant (III) suggests an emphasis on resilience, in order to adjust performance quickly and take advantage of the unexpected opportunities arising from high complexity. In quadrant (IV) Lean and resilience seem to be equally important. Whereas Lean can be useful to maintain the stability of a linear system, resilience may be useful to anticipate and control threats that could lead the system to migrate to quadrants (I) and (II). Future work will explore this proposed framework.

Conclusions

As discussed in this chapter, RE and Lean can be synergistic. On the one hand, RE may add important context to Lean implementations, as it stresses the value of performance variability as well as provides tools and theories for understanding WAD. Lean theory also supports understanding WAD, but there is anecdotal evidence that this aspect is often underestimated in Lean practice (Radnor et al., 2012). Indeed, Lean-as-imagined by theorists is likely to be often different from Lean-as-done by practitioners. Additionally, by giving visibility to the complexity of health care, RE can support the identification of those processes that are not fit to the use of Lean practices in the same way these are implemented in manufacturing plants. On the other hand, Lean production offers to RE tools and production management principles (e.g., use small batches) that may support the design of more resilient systems, as well as it provides insights into RE tools, such as the FRAM. The value of these synergistic relationships is still mostly theoretical, and both the Lean and RE communities are invited to provide the necessary empirical evidence. Those involved in such initiatives must be well versed in each theory so that they can move beyond stereotyped views. Of course, technical knowledge and theoretical alignment are not enough for a successful integration of both approaches, as a number of other factors can play a role. A key factor is the reason for using Lean and RE. For instance, if Lean is used as a cost-reduction programme with short-term goals (which is not recommended by Lean theory), concerns with understanding WAD and the impacts of complexity may be seen by management as an inconvenience. Lastly, joint use of Lean and RE in health care requires an understanding of how the barriers to the implementation of each paradigm interact. Some barriers can be unique to each paradigm, such as the fact that the qualitative emphasis of RE research is in contrast with the values of clinical research that dominate modern health care (Hollnagel, 2014b). Other barriers may be common to both Lean and RE, such as the introduction of new terminology, identified by Brandão de Souza and Pidd (2011) as a barrier to Lean in health care.

2

The Jack Spratt Problem: The Potential Downside of Lean Application in Health Care – A Threat to Safety II

Sam Sheps and Karen Cardiff

CONTENTS

Introduction

Over the past several decades there has been increasing interest in, and expenditure on, the application of Lean thinking in health care. Hospitals (e.g., Virginia Mason in Seattle [Association of American Medical Colleges; Kenney, 2011], and Flinders Medical Centre in Australia [King et al., 2006]) as well as health systems (e.g., the Province of Saskatchewan in Canada [Government of Saskatchewan], the National Health Service in the United Kingdom [Lodge and Bamford, 2008] as well as in Denmark [Laursen et al., 2003] and Sweden [Tragardh and Lindberg, 2004]), among others, have made Lean re-engineering a central component of their improvement (quality, safety and efficiency) efforts.

Evolving from Taiichi Ohno's Toyota Production System, North America embraced what was initially called Business Process Re-engineering in the 1990s (Buchanan, 1997; Packwood et al., 1998; Locock, 2003), joining longer standing improvement efforts such as 'continuous quality improvement' (CQI) and 'total quality management' (TQM). The primary aim of Lean was reducing of waste (*muda*), stabilizing uneven demand and variability in outputs (*mura*) and improving working conditions (*muri*). Lean also used new tools such as takt or cycle time, 'value stream mapping' (VSM), and 'just-in-time' (JIT) supplies management and existing measurement processes, for example, 'six sigma'. The move to Lean also included a new vocabulary (e.g., sensei, hoshin kamri) and trips to sites that had undertaken Lean, such as Virginia Mason Hospital in Seattle, Utah and Japan in what John Seddon has called "industrial tourism" (Seddon and Brand, 2008).

Why Lean?

Despite a history of quality improvement efforts in North American health care, spanning some 50 years, the issues of efficiency and effectiveness (i.e., quality) remain a challenge. In particular, efficiency, which Janice Gross Stein observed was generally considered more of an end than a means (Stein, 2001), continues to be a never-ending problem. Health care universally is challenged to deliver 'better' and more technologically complex care in a timely way to more people at lower cost. Downsizing hospitals in the 1990s was thought to be a useful response to this challenge but several predictable unintended effects occurred: acuity of patients increased, costs were shifted (not eliminated) to community settings that had to care for more complex patients and many problems with quality of care remained. The new emphasis on 'waste' in Lean thinking and methods (e.g., 'VSM') was seen as an innovative approach to these problems. Moreover it seemed logical that 'waste' – that is, care or administrative processes adding no benefit to the patient – could easily be dispensed with as a way of reducing cost and improving quality: a perfect combination.

Some Problems with Lean

Despite widespread enthusiasm, there are several issues that make one sceptical about the sudden rush to Lean. In North America while the rhetoric supporting Lean talks about quality, in practice much of the work done is really about cost cutting. One of the early applications of Lean in Saskatchewan, for example, was to reduce the cost of a new Children's Hospital in Saskatoon by $100 million, approximately 25% of the initial budget. Other examples include reducing labour costs (Esimai, 2005; Spear, 2005) and improving productivity which for 60 years has been a never-ending quest. Lean is also credited with improving patient flow in a number of settings (e.g., outpatient clinics, cancer clinics, emergency departments, reducing medication loss, improving staff morale, reducing time to diagnosis and streamlining office processes such as work flow and paperwork). Whatever quality improvements these efforts had, a key claim was reduced costs as either a direct or indirect benefit.

While creating 'value' by cutting 'waste' is a central tenet of Lean, it is not clear whose 'value' is enhanced (or 'waste' cut). Cost reductions clearly are of value to health care institutions and systems (and possibly even individual clinicians), but it is not always clear to what extent they are of 'value' to patients or families. Indeed the emphasis of Lean redesign is often limited to quantifiable operational costs and staffing ratios: the cost reduction leads to quality perspective, as opposed to the more empirically demonstrated finding that quality improvement leads to cost reductions approach (James and Savitz, 2011). Improving work flows and reducing wait time may indeed increase aspects of the quality of care and enhance patient/family satisfaction, but at a cost to them in terms of thoroughness of care: a form of efficiency–thoroughness trade-off (Hollnagel, 2009a). Young and McClean (2008) note a confusing mix of value-concepts that Rust (2013) suggests stems from "a complex and fragmented customer community". Moreover, as Waring and Bishop (2010) point out, Lean, as well as other improvement initiatives, creates areas of contested control (and therefore 'value' definition) in the struggle between the "logic of managerialism" and the "logic of professionalism". Another 'perspective' rarely identified or discussed is that of time. The time horizon applied to the process of determining what is considered 'waste' (or not) is critically important. The value of a particular process or activity might not be immediately clear, but when viewed with a longer-term perspective, with input from front-line staff who understand the effects of a particular process on downstream workflow, might be evaluated differently.

Regardless of whose 'value' is being created, the evidence in the literature remains largely anecdotal and/or describes one-off applications of Lean. There is no research that takes a system-wide perspective on the impact of Lean and little formal evaluation or analysis of effects using rigorous designs (Young and McClean, 2008; Proudlove et al., 2008; Radnor and Walley, 2008; Vest and Gamm, 2009; de Souza, 2009; Mazzocato et al., 2010; DelliFraine et al., 2010; Radnor et al., 2012; Rust, 2013). Moreover, there is an ongoing debate in some of the current literature regarding whether and how Lean is relevant to health care at all given its service as opposed to manufacturing orientation. Some processes in health care do involve supply chain management–type activities (stocking pharmacies for example), but many clearly do not. Rust (2013) for example notes that "Lean is not the appropriate method to use in markets with high demand variability and high requirements for customization, as is the case in health care …". Furthermore, "… following Lean redesign strategies may create a trade-off between service delivery efficiency and the ability to respond quickly to changes in customer demand [patient care needs]" (Rust, 2013, p. 82). Indeed exploring the problems of Lean from a supply chain management perspective, as Rust (2013) does, highlights an intriguing conjunction of ideas relevant to resilience (see below).

The application of Lean, in health care based on the inculcation of tools, targets and plans, is highly scripted ('VSM' for example) and, as noted earlier, entails learning a host of exotic words. Interestingly, with regard to tools, Taiichi Ohno did not think in these terms. He said "Write nothing down!" and that "T" was not for Toyota but for "Thinking". In effect he was talking about a culture of improvement (developed over many years), deeply ingrained and based, in part, on the input from front-line workers on how to make processes better. This is worlds away from the usual management mantra, "let's just reduce costs and solve this one problem" that reflects the basic "diagnose and treat culture" pervasive in health care.

In contrast to viewing Lean as deep cultural change, its applications in health tend to be highly focused on isolated and disparate clinical activities (processing of blood for transfusion, sterilization procedures, central intravenous line insertion procedures in the intensive care unit [ICU], re-organizing clinical laboratory or pharmacy space to reduce time and energy to carry out functions, pre-operative assessment, discharge planning, etc.). The degree to which any overall organizational learning occurs and the culture actually changes is unclear. As Radnor et al. note: "… in practice Lean became a constellation of disjointed and poorly connected activities … a technical fix for tackling pre-existing problems" (2012, p. 369).

Controlling flow of care processes (e.g., discharge planning or clinic patient movement through diagnostic testing) to reduce variability – that is, managing variance – may have the untoward effect of making the care processes too tightly coupled. They thus become brittle and unable to respond appropriately to surprise and other perturbations. In complex adaptive systems performance variability occurs for a variety of reasons, for example, the need for work-arounds of rigid procedures or policies due to contextual features of the work that make the procedure or policy unworkable. It is well recognized that resilience in complex adaptive systems requires a certain amount of slack or reserve to enable appropriate responses to uncertainty and surprise. Thus considering variability or slack as 'waste' rather than an asset can have unfortunate consequences. Moreover, because adaptation to existing conditions is an emergent property of how work gets done, tightly constraining variability forces front-line staff to engage in more rather than fewer work-arounds, compounding the problem. The 'easy' response is to attach the blame to a lack of procedural or policy compliance, thus to greater enforcement of compliance with increasingly negative effects on morale.

Lean creates contradictory perspectives regarding work-as-imagined and work-as-done (Hollnagel, 2009a, 2010; Cook, 1998; Dekker, 2011; Dekker et al., 2013). On the surface Lean is about understanding how work gets done so as to identify unnecessary steps in care processes ('waste') and to constrain variability as noted above. This is done through engagement of front-line staff that in theory is highly appropriate. However, imposition of the Lean philosophy, apparently critical to its successful implementation, may undermine getting a true picture of how staff actually manage daily work and may create contradictions that are hard to reconcile since work-as-imagined may trump work-as-done. In part, as noted above, this is an issue of whose 'value' is considered a priority. But at a more fundamental level it is more often a way of co-opting the perspectives of front-line staff to management perspectives (Waring and Bishop, 2010). Moreover, limited or no access to uncensored insights from front-line staff (necessary to understand the reality of Lean impacts on care delivery) can be seriously problematic: particularly with regard to overly optimistic assessment of Lean outcomes that are managerially preferred and/or 'politically correct'. With the loss of meaningful front-line reality, 'waste' may also be magnified.

While much is made of the importance of Lean to the bottom line, rarely is return on investment adequately assessed. In many health care contexts claims are made regarding the money saved but such estimates tend to be rather global and vague (e.g., ICU savings of $500,000 a year [Radnor et al., 2012]). Moreover, it is not always clear where this money saved actually goes.

Examples of Potential Problems with Lean Application Health Care

Because Lean is an approach to systematically reduce or eliminate 'waste' in organizational processes, while enhancing 'value' for the patient, it is not surprising that many of the efforts in health care have focused on the problem of duplicate tests, patient waits, as well as procurement of equipment and supplies and utilization of staff skills and time. Described next are several examples of the potential unanticipated consequences of Lean application in health care, particularly when context is not properly considered or ignored completely. In particular, these cases highlight the challenge of implementation.

> **Example 2.1: "Creating a Trade-Off between Service Delivery Efficiency and the Ability to Respond Quickly to Changes in Customer Demand" (Rust, 2013)**
>
> Building a new hospital was viewed as an opportunity to employ Lean to improve quality and productivity and to reduce costs in the new space. As an example, the JIT strategy was applied to the planning process for the procurement of medical supplies in the new facility. It was observed that the JIT strategy was working very effectively to reduce excess inventory of medical supplies in the institution that was being showcased as an exemplar of Lean processes. The resulting decision to almost entirely eliminate storage space for medical supplies in the proposed new facility created, as Rust (2013, p. 82) states, "a trade-off between service delivery efficiency and the ability to respond quickly to changes in customer demand".
>
> Unfortunately, in this instance, the individuals involved in the planning process did not give due consideration to context, ignoring

critical environmental differences between their site and the exemplar site (e.g., patient population, size of the community, access to supply-chain and service-delivery providers, seasonal weather challenges), which would make it virtually impossible for JIT to work effectively, as it did in the exemplar site. An extreme example of the risk of applying a JIT approach across all aspects of medical supply management regardless of patient population and specific types of devices could be of use to endotracheal tubes in a paediatric setting. Given the fact that an endotracheal tube is required immediately in life-threatening circumstances, JIT delivery is not good enough, particularly in a paediatric facility where you need endotracheal tubes of different sizes. This example highlights one of the problems noted earlier regarding controlling flow of care processes and eliminating any slack in the system with respect to quick access to important medical supplies, thus making the system brittle and unable to respond in a timely manner to surprise. It can potentially force staff to engage in more, rather than fewer, workarounds compounding the problem.

Example 2.2: Improving Access and Flow and Building Capacity in Outpatient Care and the "Whiplash Effect"

An outpatient care facility used Lean to redesign their service model for patient intake, with the intention of improving patient access and flow and improving the overall organizational capacity to increase throughput and decrease costs. The major flaw in this instance was that the facility failed to seriously consider the range of complexity across the already critically ill patient population (with both medical and psycho-social vulnerabilities) that it served. Unfortunately the outcome of the Lean redesign process was a rigid application of the new model, which eliminated 10 minutes of assessment time for the intake personnel. In some instances it was possible to complete a thorough intake assessment in the allotted time, but in many instances, it was not. This is an example of the contrast between developing a service delivery model based on work-as-imagined and failing to adequately consider work-as-done. Practitioners at subsequent stages of the care path raised serious concerns about the new model, arguing that it was not maximizing the patient experience and frequently resulted in incomplete capture and/or documentation of important patient information that created downstream duplication of effort to obtain the missing information critical to treatment (known as the 'whiplash effect') (Rust, 2013). In addition to not maximising the patient experience, time is not saved – it is simply spent during subsequent stages. Worse, in some cases treatment was delayed since the relevant patient details were not readily available for important downstream specialist decision making. This example highlights the challenge of distinguishing 'value' from 'waste' and reinforces the importance of not only obtaining staff insights about work-as-done versus work-as-imagined, but also carefully managing the process used to engage staff opinion, remaining mindful that staff can potentially be co-opted into agreeing with management. It also points out that VSM cannot ignore the perspective of time.

Example 2.3: Improving Identification of At-Risk Patients

The combination of chemotherapy and radiation therapy given at the same time is known as dual therapy, and in some cases the patients on dual therapy are at a higher risk of side effects. A cancer clinic recognized the risk of not properly identifying patients for whom the dual therapy was being given, and, engaged in a 1-week Kaizen exercise to develop a solution that would minimize or completely eliminate the risk of this happening. Because the problem was one of properly identifying patients on dual therapy, a decision was taken to label the patient's medical record with a specific image. However, it turned out that the selected image was an already existing image that carried a range of meanings to different practitioners working in the facility. The already-in-use image was meant to identify patient records that contained 'new' information about the patient, and although this could be said to be true about the patients on dual therapy, the image could actually be interpreted differently, depending on the person looking at the chart. For example, it could mean new lab results were available or the patient's medication or treatment regime was recently adjusted. The Kaizen exercise had chosen to fix a problem by using a generic already-in-use image to solve a particular risk. Eventually, the cancer facility developed an entirely new image that was meant for a specific use, i.e., to red-flag patients on dual therapy, but in the interim the Kaizen 'solution' gave practitioners a false sense of security, putting patients at risk. This example highlights the potential problem of fragmentation or the risk of identifying solutions within silos of care. The highly specialized clinic used a systematic process to engage their staff to solve a specific problem, with the objective of reducing risk to a subset of their patient population, yet things only became more complicated.

Conclusion

Despite the enthusiasm of many, the application of Lean improvement methods in health care is at best problematic, and needs to be carefully assessed, and at worst is dubious and poorly matched to the complexity (unpredictability of demand) inherent in health care. Both the reasons enumerated above and the examples suggest that greater understanding of the impact of the application of Lean in the health care context is essential. Aside from the issue of whose 'value' is really being improved, the concept of 'waste' in health care processes requires considerably deeper thought, and more attention needs to be paid to unintended effects. Primarily the issue is one of removing too much slack, leading to the creation of brittle processes that have little capacity to handle the everyday surprises that occur. Cook, after Rasmussen, has neatly described why Lean initiatives should be undertaken with caution, as they essentially move the operating point of any process closer to the marginal boundary of safety (Cook, 2013a). Thus zealous Lean is the antithesis of resilience.

An alternative approach to enhancing safety and quality that is synonymous with resilience is "agile process re-design" which is defined by Rust (2013, following Gunasekaran 1998) as "... the capability to survive and prosper in a[n] environment of continuous and unpredictable change by reacting quickly and effectively to changing markets, driven customer-designed products and services". While clearly reflecting a manufacturing context, Vries and Huijsman (2011) make a compelling case that 'agile process re-design' is relevant to health care. It is an approach that allows for considerations of how work could be done differently to increase provider or organizational productivity and patient satisfaction while retaining the focus on them to maintain the capacity to respond to demand variability and surprise. The critical notion missing in Lean concepts and applications is that allowing flexibility in how work is done creates success whether we want to call this agility or resilience.

3

Recovery to Resilience: A Patient Perspective

Carolyn Canfield

CONTENTS

Seven years ago my husband, aged 35 years, was found unresponsive in his hospital bed 8 days after a successful surgery, to everyone's surprise. During that week he had experienced many common post-operative complications. I felt that we were receiving excellent care from skilled nursing and physician specialists. How could their breezy assurances of recovery have been so wrong?

The Shock of Harm

The bedside caregivers were devastated, overcome by the unexpected death of their patient. I held them as they convulsed with sobs and reassured them that my husband had lived a wonderful life. I thanked the surgeons for their skill and care. That morning I said that I wanted to salvage something good from whatever had happened, to see that other patients and families did not experience something similar.

Although medicine and health care were unfamiliar worlds at the time, we had fully accepted there were risks with every therapeutic choice we considered. But learning that patient safety was an evolving discipline was startling. I wondered: how could a patient be unsafe while in the hands of the very expert care that we had been expected to trust? Asking again and again, "How could this be?"

Teasing a narrative from the medical chart revealed a complexity of unnoticed gaps. From inside health care I heard from everyone, "It happens every

27

day". How could routine risk not factor in everyday precautions? Why were practitioners just as startled by this death as me? Had avoidable tragedy become banal or tedious to whomever was responsible? Could caring professionals truly tolerate this?

I came to see that my husband died from a failed model of fragmented care in which different symptoms were managed by different specialties, and everyone did their job in the snapshot of a consult. No one strung those snapshots together into a moving picture of a patient, in this case, one accelerating to catastrophe. No one took responsibility for the whole patient. Gradual deterioration went unnoticed in the bustle of shift rotations and multiple orders that compounded poor continuity of care.

Curiously, this death was not identified as an adverse event. Indeed, there was no recognisable error, no discrete mistake or failure, no missed box on a checklist. Everything in the operating room had gone so well, so why look further? Surprisingly, the discharge summary on my husband's chart stated the wrong cause of death with no mention of autopsy findings unrelated to the surgery and co-morbidities. None of my husband's carers would discover why their patient died. To my astonishment, many would not even learn that he had died: "We have no mechanism for that". There would, apparently, be no learning. It was hard for me to align such system blindness with the meticulous technical expertise and devotion to caring I had observed. The trust we had placed in care felt betrayed.

A System That Fails Everyone

To enter professional practice in medicine requires an immense drive to help, to heal and to offer hope, and thereby to dedicate one's career to caring. But entering medicine also imposes a daily work environment where cultural hierarchies of status, authority and respect may inhibit trust and mutual support among and within professions (Klass, 2007). As medicine has become more specialised, urgent and vital clinical tasks rely on highly trained, expert interdisciplinary teams to pursue more complex and riskier challenges with high expectations for success (Groopman, 2007). When an unanticipated failure in care occurs, and when its impact on practitioners is ignored, the isolating trauma can produce guilt as well as damage professional self-esteem and dedication (Banja, 2004). So I asked myself: is it possible to design a more perfect system for burnout? I needed to understand how good people could be so unaware and so vulnerable?

Rarely does training equip clinicians with the self-assurance, humility or even language to approach a harmed patient or grieving family to offer an explanation for whatever has happened and help them to seek understanding (Iedema et al., 2011). Offering an apology in exchange for forgiveness may be the ritual through which we can best overcome trauma (Lazare, 2004). This positions the professional to re-establish mutual trust and rebuild confidence for both the provider and harm victim. But ease with disclosure and support is unusual and often discouraged, in spite of the harm its absence imposes on practitioners and patients alike. Indeed there are also apparent gaps in support services for those who are involved in medical error and are under investigation for their role in the events (Madhok et al., 2014).

Medical insurers seem reluctant to accept the evidence that an explanation and apology dramatically reduce the likelihood of litigation (MacDonald and Attaran, 2009). Plainly, patients sue for information and to re-establish trust. Thus reducing medico-legal risk exposure might be better addressed with openness for a more rapid recovery to full functioning for all harm victims, including the patient, the family, clinicians and the wider organisation (Iedema et al., 2011). Instead, insecure provider organisations seem to close down, deepening and prolonging harm to all involved. Dysfunction continues with degraded communications between less trusting workers and fearful care recipients. Might such system brittleness invite disaster – individually and collectively?

An insidious blame-and-shame culture with inadequate professional support would exact a substantial toll on those who spent long years of training aiming to be perfect. Who could afford to risk freely investing heartfelt care in such a hazardous and punishing environment? Personal reward from choosing a "caring profession" may drain away with isolation smothering solo efforts to recover. For some the pressure is too much, leading clinicians to leave their practice or even take their own lives (Banja, 2004; Madhok et al., 2014). Perhaps practitioners lose their willingness to care, trust and contribute—their capacities that are so critical to system-wide resilience.

A hierarchical medical tradition has evolved through the twentieth century into what often looks like a punitive culture perpetuated by allegiance without trust (Sharpe and Faden, 1998). A closed, even secretive, climate judges care failure as professional failure without a path to redemption (Berwick, 2009). Such censure also evades honest disclosure and accountability to harmed patients and fellow clinicians. But change is underway. Autocratic leaders, disempowered staff and passive patients are rapidly giving way to roles guided by higher expectations (Hurwitz and Sheikh, 2009). In this era, a new style of focused multidisciplinary teamwork is demanded, including highly individualised and open relationships that are built on trust throughout the network that makes up health care: among practitioner teams, between patients and their support communities, and among governors, managers, professional organisations and the public.

Networks of Trust Defining Care Relationships

Rarely these days does health care function as the stereotypical doctor-directing-patient set piece, where authority is predictably asymmetrical, with the best patient as the most compliant one. Today, an infinitely complex web of relationships requires trust to transcend geography, time, clinical setting and task across the multiple dimensions and accelerated pace of modern service delivery (Hollnagel et al., 2013). Trust must often be presumed with no time to weigh individually the merits of each dependency, care complexity is so massive.

With unending layers of complexity constantly moving, comprehensive mapping of health care flow is daunting. Given the abstraction required for much process analysis, proposed improvements might very likely address how to alter work-as-imagined, rather than work-as-done (Hollnagel et al., 2008). A full understanding of work-as-done remains incomplete unless all actors are able to participate and the diversity of their individual experiences is welcomed. Even then, the complex adaptive system of health care will likely have morphed into something slightly different an instant later, and then again and again. It may well be that the culture surrounding practice is perpetuated by a fantasy world of work-as-imagined, with little reality underpinning our assumptions and expectations of others and for ourselves in the relationships that define health care delivery.

An unexpected adverse medical outcome may shatter relationships that are based on false assumptions and assumed commitments (Frank, 2004). Such shocks can destabilise trust between the practitioner and patient, as well as among team members who must rely on each other to carry out medical tasks, to direct a care plan or to hold the system accountable to the public for maintaining high-quality service. A rapid response to repair destabilised relationships following a mishap should include a protocol to inform victims, to explain events, to apologise, to support recovery and to undertake a system remedy. Routine practice of these elements would go far to re-establish and even strengthen the trust that is so fundamental to system functioning.

Should those challenges to patient/family confidence in health care remain unsettled, doubt can give way to the most enduring of harms: a betrayal of trust (Walker, 2006). The longer the ritual of giving account and apologising is delayed, the more easily the experience of betrayal can deepen, then transform and spread to poison future confidence in relationships and in medicine as a profession or as a service. What starts with an unmet individual need for closure may rapidly become a catastrophic failure, difficult or impossible to contain, such as widespread public outrage over a poorly managed and highly publicised scandal, or the tragic resonance of a professional suicide.

Long after surgical site infections resolve or medication errors are managed, the sting of disrespect and abandonment may interfere with relationships essential to bonding patients with carers, practitioners within teams and public governance bodies (senior health care managers, boards and ministries of health) with their constituents (Pilgrim et al., 2011). The erosion of trust reduces options for resolution and the creative scope that extensive networks of respect and confidence can foster. Scarcity of trust may limit the possibility of accommodating both system and personal stress successfully. Inadequate coping with stress reveals an overall system becoming increasingly brittle and intolerant of shocks, jeopardising safety and increasing the likelihood of new harm.

With a rising priority placed on health care effectiveness and efficiency, new methodologies are proliferating to measure the dimensions of trust and their strength within provider teams (Gittell, 2009). The best approaches to analysing interrelationships attempt to explore the nuances of work-as-done, in contrast to the expectations of work-as-imagined on which training and safety systems often rely. Understanding how teams work best will reveal factors that reinforce trust and enhance agility for safer and more efficient care. Communications, role clarity, skills coordination, stress sensitivity and collegial respect can all foster care excellence through stronger team relationships. Trust may have a tendency to spread just as effectively as mistrust (Frank, 2004).

However, notably missing from many studies of teamwork is consideration of patients and families as influential and independent members of care teams. Patient safety and practice standardisation seem to have evolved from a system-centric view of work-as-imagined. For any model of care that purports to be patient centred, it seems essential to account for the dynamic trust that guides patients through care experiences. No episode of a therapeutic intervention can be analysed comprehensively as work-as-done without considering the unique interactions of the patient and the patient's network of support in the community.

Citizen-Patients Transforming Health Care

Encountering a system insufficiently sensitive to the harm it creates, and seemingly unable to learn from tragedy, provoked my sense of the possible. Could recasting the role of the patient through creative and productive reform lead to greater sharing of authority with those who have the most at stake for outcome (safety) and the greatest knowledge of care delivery (quality)? I soon took notice of intriguing talk about an evolving "democratisation of health care".* But who might lead change in a social movement to redefine respect for the patient across the care continuum?

---

* Personal conversation with Nigel Murray, former chief executive officer (CEO) Fraser Health Authority, *Medical Makeover: Redesigning Physician Services for Tomorrow's Health System,* September 27–28, 2011.

Imagine a role for the citizen-patient that allows a member of the general public to choose to join with health care professionals to improve care at a population level by representing a missing perspective on work-as-done. The rewards in altruism for benefiting the broad sweep of the public interest combine neatly with a personal benefit in living in a society served by more equitable, cost-efficient, easily accessible, higher-quality and safer health care, grounded in evidence and respecting individual values and preferences. In Canada popular activism in health care improvement seems patriotic, so sentimental are we in our pride for a single-payer health care system that distinguishes us from the Americans. And of course, the citizen-patient can hope that an improved system is ready to meet their expectations in their own time of need.

The *British Medical Journal* proclaimed in its May 14, 2013, editorial, "Let the Patient Revolution Begin" to usher in a new era of democratised health care and the grass-roots reform necessary to achieve it: "… how better to [improve health care] than to enlist the help of those whom the system is supposed to serve—patients? Far more than clinicians, patients understand the realities of their condition, the impact of disease and its treatment on their lives, and how services could be better designed to help them" (Richards et al., 2013).

Through middle age and beyond, we baby boomers are almost certainly exposed to unfulfilled expectations for excellence in care for family and friends, even ourselves, as an uncomfortable clash between work-as-imagined and work-as-done. Whether our expectations are unrealistic, or the system is actually failing, is less important than acting with courage to support social momentum for a new role. Indignation is fuelling yet another wave of this generation's inclination to consider social change as a natural cultural expression.

Once allowed inside a very closed system, the public will discover, as I did, a sincere commitment to improvement, but solely through the eyes of the provider, not the recipient of care. A growing pool of volunteers is seeking to align their aspirations as system users with those of clinicians, managers, researchers, educators, policy leaders and improvement theorists. Newly retired with discretionary time and equipped with a wealth of career-honed skills, volunteers in health care reform can offer expertise as patients and carers to benefit projects and processes that would otherwise lack their user perspective and energy.

In many jurisdictions, public agencies are welcoming such public willingness by developing engagement methodologies to join up patients and the public with the health care sector. With coaching and support, experiments in collaborative design, implementation and evaluation are building confidence for new forms of involvement. Health care clients and members of their community networks can participate as full-status team members in improvement projects with productive and often novel strategies for impressive results (Spencer et al., 2013). Co-design of services and co-production to embed improvement is rapidly moving from tentative trials into mainstream practice.

Collaboratives between health care providers and public volunteers can take advantage of information technology to involve broad participation spanning vast distances in a diverse geography at modest cost. Some hard-to-reach populations may be able to participate only through electronic connectivity to surmount physical accessibility and communication limitations. Taking consultation and participation to marginalised populations has demonstrated the broad willingness of the public to contribute their insights, articulate their needs and identify solutions. Citizen-patients can develop their skills and confidence through discussion boards, Web-based seminars, teleconferences, self-guided training and resources for research. Learning from experienced peers may be among the most valuable and efficient training delivery model.

Clinicians and project leaders can simultaneously develop their capacity for citizen-patient collaboration through mutual and collaborative education: the team that works together should train together. Policy, evaluation and governance can develop authentic and thorough public accountability by inviting patients, carers and the public as productive equals at the table. Given the will and imagination, experience-based collaboration with citizen-patients can rapidly become feasible in every sector of health care.

A rapid growth of engagement with care users will meet head on the challenge of representative patients and care communities. Because current mechanisms for participation usually stem from work-as-imagined, citizen-patient leaders hold a special responsibility to urge development of successful engagement methodologies that validate the experience of all service users in work-as-done (Schubert et al., 2015). It will take a proliferation of models to engage the diversity of voices necessary to build strength, confidence and trust throughout the network of relationships that defines health care as experienced.

With progress in patient- and family-centred care, health care providers are increasingly inviting patients into collaborative shared decision-making for treatment and services (Edwards and Elwyn, 2009). For some patients, but not all, this will become a welcome new role as co-lead for their own care planning, including choosing a preferred care delivery setting and assessing care quality. Other patients with limited and diminishing capacity would consider this an unappealing or impossible burden of treatment (May et al., 2014). The intent is to ensure that patients receive the care they need and want, when and how they need and want it (Berwick, 2009). Meeting an individualised ability and desire to self-direct and self-manage care, with access to sufficient infrastructure, capacity and resources to sustain such independence, is a growing challenge for public policy as well as operational practice. But there can be no return to the authoritative physician and subservient, compliant patient in a "take it or leave it" care plan.

Co-creation of medical care services has the power to transform patient and professional roles. The recipient of care, as well as the family/neighbour/friend supporter of care, becomes an incontestable authority on care delivery as work-as-done. A greater role for the patient may strengthen accountability and coordination among health care professionals themselves, as it is only the patient's view that sees the complex of interdependence across the care system. Persistence in navigating one's own care path builds unique and expert knowledge from experience to equip the patient to help redesign the system as a valued authority in what works, what does not work and what might work better.

Much remains for the effective and practical integration of the patient experience into system reform strategies. Productive engagement techniques depend on a better ethnographic understanding of health care processes that include the patient as a dynamic player. A recent attempt to identify hazards and control risk discussed in "Safer Clinical Systems: Evaluation Findings" from the United Kingdom tentatively involved patients to clarify work-as-done and identify strategies to reduce risk (Dixon-Woods et al., 2014). A lack of adequate engagement methodology was found to limit project achievements. These shortcomings should not discourage attempts to access the insights of patients, perhaps by tapping their very creativity and ingenuity directly by asking them to suggest better ways to engage a wealth of experience.

Continuing research might consider: How does patient-led care fit with (or challenge) our concepts of patient and professional safety? What do patients and their families need to learn and practice to succeed as patient safety experts? Does a greater patient and public awareness of hazard create new vulnerabilities that might disrupt care and challenge resilience? The needs and capacity of all patients are not alike, so can we develop sufficient sensitivity to that diversity when we assess system safety?

Democratised Health Care as Resilient Health Care

Without inviting the system user's experience to define, observe and report safety and harm, there can be little meaning to resilience. As a start, the Resilience Analysis Grid (Hollnagel et al., 2011) offers four dimensions for examining the potential for citizen-patient contributions and opportunities for safer and more successful health care under an increasingly democratised model of health care. A resilient system has the capacity to respond, to monitor, to anticipate and to learn from experiences in every setting across the health care continuum. Inviting past, present and future patients and their carers to grow from disempowered outsiders to become active participants will enhance health care's need to comprehend quickly, flexibly and sensitively those components of care that only care recipients can perceive.

A genuine examination of health care resilience must engage the entire network of reciprocal relationships of trust that comprise care provision.

For each of the dimensions of resilience, exploring the character of trust rela-tionships might be guided by the following initial inquiries.

- **The ability to respond:** *How ready is the organisation to respond when something unexpected happens?* Do health care workers explore with patients and family members what safety means to them? What is the full range of safety precautions? Do providers respond effec-tively and proactively to meet patient, practitioner and public needs following harm, or the risk of harm, before harm's impact escalates?

- **The ability to monitor:** *How well is the organisation able to detect changes that may affect its operations?* Can patients freely inform pro-viders when they feel safe, or unsafe, or vulnerable to harm? Would providers listen and react if patients experience a change? Are confi-dent and trusting patients considered and assessed as a system goal? Do confident and trusting patients also experience better clinical outcomes that maximise value for care delivered?

- **The ability to anticipate:** *How large an effort does the organisation put into foreseeing what may happen in the future?* Are citizen-patient advi-sors embedded in improvement processes, planning and evaluation? Are patient peer counsellors distributed throughout care settings to support patients and families, observe care effectiveness, detect change and play an early warning function for the impact of system stresses on clinical performance?

- **The ability to learn:** *How well does the organisation use opportunities to learn from what happened in the past?* Are patients and family mem-bers integrated into safety reviews and system resilience scanning? Do patients and family members confidently provide specific, timely and detailed feedback to providers on drivers of care excellence as well as near-misses, harmful incidents and their impacts? Are patients and family members encouraged to propose innovations and improvements for secure and trustworthy care experiences?

In health care, personal resilience is essential for system resilience. Reflecting on the tragedy that compelled me to learn about health care dynamics reinforces my conviction that only when all are present, participat-ing and accountable for our part will we gain a true and useful understand-ing of what is safe and what is unsafe:

> Safety is the system property that is necessary and sufficient to ensure that the number of events that could be harmful to workers, the public, or the environment is acceptably low. (Besnard and Hollnagel, 2014)

Trust relationships are ever more important as health care systems are asked to do far more with limited resources to achieve better outcomes and sustain safety for all. Only if everyone can collaborate with a shared understanding and respect for the internal values and vulnerabilities that connect us all can we understand what resilience means to the human experiences in health care, those of dignity, suffering, survival, humility and relief that underlie this dispassionate recitation:

> A system is said to be resilient if it can adjust its functioning prior to, during, or following changes and disturbances, and thereby sustain required operations under both expected and unexpected conditions. (Hollnagel, 2014b)

The southern African concept of 'ubuntu' seems most apt in characterising the promise of a democratised health care system where the fully engaged citizen-patient embodies collective resilience for all, as a matter of trust: *I am what I am because of who we all are.*[*]

Acknowledgement

The author would like to acknowledge the exceptional support of Dr. Sam Sheps at the School for Population and Public Health, University of British Columbia, for introducing me to resilience and for his generous assistance in the preparation of this chapter.

---

[*] Ubuntu. The Ubuntu story. http://www.ubuntu.com/about/about-ubuntu.

4

Is System Resilience Maintained at the Expense of Individual Resilience?

Anne-Sophie Nyssen and Pierre Bérastégui

CONTENTS

Introduction

The theme of this book (work-as-imagined and work-as-done) seems very close to the distinction provided by Leplat between task (work as prescribed) and activity (work as done) at the French Language Ergonomics Society (Société d'ergonomie de langue française, SELF) in the 1970s (Leplat, 1975, 1990). It may be interesting to trace briefly the reasons French ergonomists had for distinguishing *task* and *activity* at this time, and compare them to the recent resilient health care (RHC) trends. The cultural diffusion of a given idea does not occur at random in a society: rather, along the lines proposed by Thomas Kuhn (1962) when describing the evolution of scientific ideas, part of the world must be 'philosophically' ready to receive the new paradigm. The origins of the Francophone ergonomics movement lie in the social issues and labour conflicts associated with it. It was an attempt to develop a scientific analysis of how the work environment and its constraints influence individuals' behaviour and health. Their social context was the unprecedented growth of mining and industry, and the devastating collateral effect on workers' conditions and health of the then-dominant Taylorist model of 'scientific work organisation'.

BOX 4.1   ADAPTATION LOOPS

All organisations define specific goals and resources and, consequently, constraints and boundaries for human performance. This work environment, with its prescribed goals and constraints, influences operators' behaviour, which, in turn, produces effects at both the systemic and individual levels. When the actual effects do not conform to expectations, due to external or internal variability, a primary regulatory loop (loop 1) intervenes, in which the operator adjusts his or her behaviour to the difference observed, at some physical and mental cost (fatigue, stress, burnout, etc.). The description of this day-to-day adaptation loop constitutes the first step of the work analysis. If, in spite of this adaptation, discrepancies persist, a second regulation loop (loop 2) becomes necessary in order to extend the interaction domain and avoid rupturing the equilibrium of the system. This loop represents the essence of French Ergonomics, which combines diagnosis and intervention in order to question goals and constraints, reduce the distance between real and prescribed work and enhance work adaptation to people.

The 'work analysis' methodology that they developed was firmly rooted in field studies. They wanted to combine understanding and acting, a diagnosis process and a change process, with the idea that both are inseparable, in order to reduce the distance between work as prescribed (what they called the 'task') and work as performed (what they called the 'activity'). Their visionary intuition was that work was a double-loop process of adapting to constraints and variability (Ombredane and Faverge, 1955), as developed in Box 4.1. However, the 'paradigm shift' they introduced was to seek mainly to adapt work to people, rather than the opposite, which was currently the case at that time.

The concept of resilience that aims to understand adaptation and adaptability and to manage everyday or exceptional variability (Hollnagel et al., 2006) could thus be seen as an extension of the SELF line of investigation. But over the course of 50 years, we might expect the concept to have evolved; so what is new or different?

Social and Psychological Dimension of Resilience

Interestingly, the social and political dimension clearly claimed by the SELF does not appear to be present in the approach to resilience as proposed by the Resilience Engineering movement. Its aim is not to adapt work to people in order to make it both more efficient and reliable, and less destructive. It is to make the system tolerant of, or resistant to, disruption. The price that actors in the system will have to pay for this 'resilience' is taken into account only marginally, if at all. And the possibility that system resilience is obtained and maintained at the expense of individual resilience is not really raised.

However, although the economic and social contexts have changed considerably, they have improved and deteriorated in equal measure for operators: fewer constraints and physical risks, but more psychosocial constraints and risks. Hence the issue of the relationship between the conditions under which the system performs and the consequences of this upon its actors, particularly those low down in the system, remains relevant.

The current 'methods' concerning business management (continuous quality improvement, Lean, faster-better-cheaper) all have the effect of pushing systems to the extreme and rendering them fragile. They also increase constraints and contraindications for employees, and lead to high levels of stress, burnout and even suicide. From a psychopathological point of view, it is the fact that these contraindications between the demands on the worker and on the organisation are not debated within the system, and rather denied, which creates intrapsychic conflict and suffering. The feeling of no longer being able to correctly perform one's work therefore increases, while societal and organisational demands for quality increase and team regulation weakens.

What has become, and what will become, of the operating methods promoted by a new RHC approach? Is systemic resilience obtained by maintaining staff personal equilibrium or is it on the contrary obtained, ultimately, by the detriment of their health, as suggested by the 'explosion' in the number of workers suffering from burnout as a result of their commitment, devotion, daily adrenaline and everyday heroism?

It is a bit strange that this question has not been asked, despite the fact that the concept of resilience has been applied to the individual psyche well before being applied to sociotechnical systems, and that its relationship to suffering during disruption and its ability to transcend traumas was broadly developed. Cyrulnik (1999) uses a metaphor for this form of resilience, which may be useful here: oysters produce precious pearls in response to injuries caused by grains of sand, which they then cover with nacre (this is resilience). But although this constitutes resilience for the oyster (the 'attack' is effectively contained), the pearl is only 'precious' for the system (i.e., humans), which involves sacrificing the oyster.

Taking this ecosystem perspective, we wanted to analyse resilience in sociotechnical systems not only at different levels (individual, collective and system levels) but also on a temporal dimension (short-, medium- and long-term consequences of the continual adaptation processes). What is the price actors in the system pay for the 'system resilience' at these different time frames? One way of addressing the subject could be to ask relevant actors themselves what they feel about 'operational crises' or emergency situations where they need to respond to major disruptions (and thus demonstrate 'individual' resilience).

Application

Following this approach, we questioned 10 emergency doctors and 10 experienced police officers about crisis situations they had experienced, the resilience resources they used to deal with the situations and how they adapted and overcame them (Bérastégui, 2014). In addition, we explored trauma using two questionnaires: the Traumaq (Damiani and Pereira-Fradin, 2006) to assess the intensity of trauma experienced and the Brief Cope Inventory to identify coping strategies used by individuals to handle stress (Carver, 1997).

Resilience and the 'Coping' Concept

Coping ability is fairly close to the concept of individual resilience. It has been widely developed, among others, by Lazarus and Folkman (1984). These authors conclude that anyone facing a stressful situation uses two complementary types of subjective evaluation: the first is to analyse the situation, its challenges and dangers, while the second aims to evaluate the available resources (internal and external) required to manage the situation. The practical actions subsequently implemented by an individual correspond to coping strategies. Two broad groups of coping strategy can be distinguished. Strategies based on the (stressful) situation itself, which aim to reduce demand or increase resources (e.g., changing goals, seeking information, drafting an action plan, seeking help) and strategies based on emotions, which aim to regulate the emotional response to the situation (minimisation, positive re-interpretation, avoidance, seeking social emotional support, self-incrimination). The relative effectiveness of a coping strategy depends on the characteristics of both the situation and the individual. For example, a situation over which the individual has little control will be better dealt with by adaptation strategies focusing on emotion. Conversely, strategies focusing on the problem will be more effective when they are used to deal with situations that are more controllable (Lazarus and Folkman, 1984). Mikulincer and Solomon (1989) demonstrated that emotion-centred coping is positively correlated with the intensity of post-traumatic stress disorder in soldiers, in contrast to coping strategies that focus on the problem.

As is the case with coping, resilience uses conscious and/or unconscious adaptation strategies. However, the idea of growth and learning differentiates resilience and coping (Anaut, 2003 quoted by Michallet, 2009). Resilience does not only mean resilience to destabilising events: it involves the individual in a learning process incorporating the development of resources and a plan.

Findings

In total, 34 traumatic situations were reported in our study by the two populations (doctors and policemen) from a few days ago to 15 years ago. A content analysis was performed in order to describe what subjects considered a critical element. Altogether 218 critical elements were identified in connection with these situations, which were then classified into more general factors. In general, crises were characterised in similar ways by the two populations. With the exception of inter- and intra-team coordination difficulties, which were mentioned only by the emergency doctors, the determining factors of a crisis were unpredictability (connected to the environment and the dynamics of the situation), emotional content (caused notably by the victims, their suffering, any perceived danger and fear) and operational difficulties (connected with, in particular, environmental constraints).

The reported resilient competencies were classified into seven main categories presented in Table 4.1. The results show that the response to a crisis depends partly upon the occupation, although in both cases the operators simultaneously used efficiency strategies to regain control over the situation (in other words, system-level resilient strategies), and coping strategies to reduce the emotional impact upon themselves (in other words, individual-level resilient strategies). In particular, the ability to take 'emotional distance' was cited as important in both cases. This competence refers to the workers' ability to protect themselves from the emotion associated with the crisis and avoid developing emotional bonds with the victim in order to be able to successfully perform the intended operational actions. The 'relational flexibility', or the capacity to create an adapted interpersonal relation with the different implicated people in the crisis taking into account the social, cultural and emotional dimensions of the situation, is only mentioned by police as a resilient competency. The other five skills mentioned are classic and can be found in the training programmes for crisis resources management (CRM) (Prince and Salas, 1993) and emergency CRM (Reznek et al., 2003).

Analysis of the Traumaq scores showed that two emergency doctors out of ten and one of the police officers presented post-traumatic stress disorder. Table 4.2 presents the individual coping strategies used by the two populations in rank order (from the most to the least used) and their correlation with the trauma score. Results showed that doctors used more emotional expression and blaming coping strategies than police officers. We also found

TABLE 4.1

Reported Resilient Competencies Deployed to Address the Disturbance

Categories	Emergency Doctors (*n*)	Police Officers (*n*)
Emotional distancing	7	7
Communication – information management	5	7
Planning, preparation – prioritising	6	9
Situation awareness	6	9
Re-interpretation – switching strategies	4	5
Leadership	7	0
Relational flexibility	0	5

TABLE 4.2

Coping Strategies in Order of Frequency (from the Most Used to the Least) and Correlation with Traumaq Score

Emergency Doctors	P Value with Trauma Score	Police Officers	P Value with Trauma Score
Acceptance	–4.735	Acceptance	.4499
Active coping	.0984	Active coping	–.1052
Planning	.1913	Positive reinterpretation	–.0884
Positive reinterpretation	–.4341	Planning	.4419
Focus on and venting emotion	.1460	Distraction	–.0924
Self-blame	.4644	Focus on and venting emotion	.4654
Use of instrumental social support	.8002[a]	Use of instrumental social support	–.2756
Distraction	.6587[a]	Humour	–.2209
Use of emotional social support	.7213[a]	Self-blame	.0907
Denial	.6096	Use of religion	.3492
Humour	.1310	Use of emotional social support	.3381
Substance use	.4656	Denial	.5119
Mental and behavioural disengagement	.1746	Substance use	.5238
Use of religion	.4656	Mental and behavioural disengagement	–

[a] .05.

a significant correlation for doctors between trauma and the social support strategy, whether this be instrumental (.8002, $p < .00$) or emotional (.7213, $p = .02$), and between trauma and the distraction strategy (.6587, $p = .04$). These strategies are associated with a high trauma score. Finally, only one inter-population difference appears to be significant. It relates to instrumental social support ($p = .02$), which correlates with a low trauma score in police officers but a high trauma score in emergency doctors. Results showed that the four most used coping strategies by the two populations have a different relationship with the Traumaq score (positive/negative), suggesting the influence of the work environment on individual resilience.

Implications for RHC Research and Practice

What can be drawn from these exploratory results in terms of the resilience process and its cost for individual actors?

Initially, there are general resilience resources, and resources that are specific to crisis situations, which need to be identified based on analysis of the situation and previous experiences. The difficulty comes from the fact that performance in context is difficult to access, although this is not necessarily negative. Moreover, resilience is not absolute; it appears to develop from a series of interactions that evolve over time between the capacities that an individual uses to demonstrate resilience and the environmental capacities that promote (or do not promote) resilience in the short, medium and long term. Thus, in our study, emotional distance appears to be a critical individual resilience skill used by the two populations during situations of crisis. However, it appears to have different consequences in the medium to long term, depending on the working environment, progressively producing psychological distress or resilience. One possible explanation is that the emotional distance, which enables individual resilience in the short term, in the long term leads to psychological damage in the empathy of emergency doctors – even though this is desired, imagined, demanded and ultimately prescribed in a hospital by the doctors themselves. This idea can be found in M.-A. Dujarier (2006), 'L'idéal au travail', which observed work in a public geriatric service. Interpersonal relationships, commitment and emotion are at the heart of the work of carers who support people at the end of their lives. Carers and also employers say they are looking for people who have a strong desire, or need, to give. The ideal thus becomes the prescription of the organisation, and people then come forward in response to this prescription. But, by considering the ideal as normal, any work conducted is constantly insufficient. It can therefore be asked whether it is possible to speak of 'emotional distance' in organisations that confuse ideal work and real work. The outcome of this discussion is not conducive to investigating the end result of the activity or the cost of the 'process' resilience maintained daily by caregivers at the expense of 'individual' resilience.

A New View on CRM and M&M Meetings

It may then be asked how best to prepare, enhance or construct RHC. Currently, implementation remains at an embryonic stage in organisations, despite the research on resilience factors.

CRM (Crew Resource Management) training developed in the 1980s in the aviation industry in response to a series of catastrophes, and could be considered as one example of an attempt to better prepare for the unexpected. This training aims to develop what is known as 'non-technical competencies', i.e., dealing with situations by making best use of available resources. It concentrates on the distribution of roles and responsibilities, managing workload, communication and sharing information, collective decision-making, leadership, the use of procedures, awareness of the situation and managing uncertainty. This type of training now appears in the medical world through the increasing use of simulation. But it can be questioned what the real benefits of this training are in terms of increasing resilience. Indeed, this training is not oriented towards management of the fundamental surprise (the 'startle effect') and the emotional content of the situation, rather it seeks to identify the difficulties people face when managing resources in crisis situations in order to reduce deviations from the 'best answers'. Recently, an awareness of this issue has led to work to develop different training. One of the fundamental difficulties of this work is the paradox of wanting to prepare for not being prepared. The general tendency of this kind of training is actually to increase the repertoire of the simulated situations and extend it to rare, unexpected situations, for which appropriate answers are prepared, and to lead the operators to follow them. One difficulty in the use of simulators as a training tool for this issue is the design of truly representative scenarios of unexpected situations. Moreover, how can one ensure the design of total surprise when people expect that something will happen to them when they are in a simulator? Part of the answer lies in the creation of a much larger events repertoire than in the past, which are randomly combined to increase uncertainty about what will happen. But the real difference is in the learning outcome we want to produce. Resilience, as we have seen, is not absolute. It is rooted in the interactions between the person's resources and the environment's capacities. There is a need, therefore, to break away from the acquisition of additional procedural competencies which cover the area of the predictable, in order to acquire general adaptive competencies. These competencies fall into three groups:

- The ability to be flexible, enabling individuals to recognise the disruption, see the need to plan other outcomes and develop new internal and external resources
- 'Sense-making' competencies that are sufficiently abstract and generic to enable individuals to rapidly judge the nature of the situation, its major risks, the things which absolutely must not be done and those that absolutely must be done, in order to regain a certain level of control without having a detailed understanding of what has happened
- Collective coping mechanisms in order to retain group solidarity and to take sufficient emotional distance to limit the negative effects of stress on cognitive capacities and cooperation in the medium and long term

In the medical world, morbidity and mortality (M&M) meetings could be considered as another example of applying the concept of resilience if and only if discussions are not limited to identifying discrepancies between actual practice and prescribed practice. Historically, M&M meetings are defined as a collective, retrospective and systemic analysis of cases marked by the occurrence of a death, a complication or an event that may have caused harm to a patient, and which aim to implement and follow actions to improve patient treatment and the safety of care (Nyssen et al., 2004). This consists of bottom-up style feedback, i.e., initiated and organised by doctors themselves on a local level with a view to improving practice and preventing iatrogenic complications. In the majority of health care sectors, there are several 'correct' ways to treat patients; a variety of standards overlap and doctors have to choose, decide, adapt and invent depending on the risks, circumstances and context of the patient. This flexibility is an important aspect of the medical profession. In such a system, feedback and failure analysis cannot be limited to measuring discrepancies and their consequences. On the contrary, it must create the context for a collective discussion on practices and guarantee freedom to 'deviate' from the guidelines, to the extent that this attitude is justified within the group, given the circumstances. In a recent study (Nyssen et al., 2012), we questioned doctors participating in M&M meetings in three different establishments on the strengths and weaknesses of such approaches. Organisational learning, which at first glance appears to be of secondary importance to the main objective of transparency taken by top-down approaches, appeared as an essential aspect of M&M meetings. Preventive action and action to improve safety are recognised as strengths of the M&M meetings by both junior and senior doctors. However, it was also noted that, for doctors, M&M meetings play a role that is just as important in terms of social support after an iatrogenic complication if and only if group discussions take place within a context of mutual trust, without judgement and where participants contribute on an equal footing. Individual experiences of encountering difficulties may transfer to the team level and then onto the organisation as a result of meetings and the establishment of improved behaviour. The process of social and dynamic interaction is at the heart of this approach and it may be asked whether this, in some way, illustrates the process of resilience in practice.

Conclusions

The pioneers of ergonomics in the 1960s saw their analysis of work as a means of putting work into question; an analysis of activity as a means of revealing the social nature of work. The current trend in research on resilience sheds new light on this work and upon the continual process of regulation, adaptation and influence between individuals and organisations. However, it must not be overlooked that resilience has a cost and, as Cyrulnik indicated, even if it is overcome, the damage remains. Today, the great challenge for resilience research is to better understand the cost of resilience on the system, the team and the individual, and how these inter-relate. It is these inter-relationships which should be considered. 'How can the system, the group, benefit from individual resilience and vice versa?'

Resilience is built on a daily basis from these interactions between protective and risk factors of the environment and those of the individual, in a Piaget's constructivism perspective (1967), as previously suggested (Nyssen, 2011). Although this dynamic can be represented, it is difficult to export and apply generic models of the concept of resilience within organisations. Rather, an RHC approach should now encourage and support a movement of self-reflection, self-regulation and self-determination, which questions ideal work, prescribed work, real work and creative work, and which perhaps will make it possible to improve resilience.

Acknowledgements

We thank the police officers and the doctors (from the emergency department of the CHU of Liège) who agreed to participate in this study.

5

Challenges in Implementing Resilient Health Care

Sheuwen Chuang and Erik Hollnagel

CONTENTS

Introduction

Despite a long-nourished suspicion that there is something fundamentally dissatisfying with the ways health care services go about improving patient safety, the established approaches and the established ways of thinking still adhere to a Safety-I perspective (Hollnagel et al., 2013). The response to adverse outcomes is typically a search for clear and simple causes, and the response to the lack of progress in patient safety is an increased emphasis on standardisation and automation. But since these responses so far have failed to bring about the desired changes, the search for alternative approaches is slowly gaining momentum. At present, the most promising alternative is the adaptation of concepts and methods from resilience engineering to health care issues, referred to as resilient health care (RHC).

Resilience can briefly be defined as the ability of a system to sustain required performance under expected and unexpected conditions. Although most of the resilience engineering literature so far has focused on the potential for resilient performance on the level of a system or an organisation, a number of researchers have pointed out that individual resilience makes a large but mostly overlooked contribution to patient safety (Fairbanks et al., 2014; Hollnagel et al., 2013; Sheps et al., 2011). This can clearly be seen in the study of how the inevitable discrepancy between work-as-imagined (WAI) and work-as-done (WAD) affects everyday clinical work and how it is dealt with by health care practitioners. Because the nature of the problems found in patient safety is the same as in other domains and in industrial safety in general, it may be useful to look at some of the methods and approaches that have been developed within the general discipline of resilience engineering. One example of that is use of stress–strain plots to make it easier to recognise and manage transitions (Woods and Wreathall, 2008); another is use of the resilience analysis grid to monitor and manage improvements, interventions and changes (Hollnagel, 2011b); and a third is the use of the functional resonance analysis method (FRAM) to model complex sociotechnical systems (Hollnagel, 2012a). The FRAM has already been used successfully in health care (see for instance Chapters 1 and 6, or Laugaland and Aase, 2014). There are also examples of methods or models proposed especially to push the trend of RHC, for instance the use of a resilience markers framework to systematically observe concrete manifestations of resilience within and across domains (Furniss et al., 2011).

Even while these experiences are useful as an illustration of how the ideas of resilience engineering and RHC can be used in practice, health care organisations still face numerous challenges and obstacles in trying to implement RHC concepts and reconcile WAI and WAD. This chapter presents an example of how some of these challenges were addressed in applying RHC in the intensive care units (ICUs) of a Taiwan regional hospital (Chuang and Wears, 2015). The study was concerned with a strategic framework to integrate Safety-I and Safety-II perspectives in the implementation of a central line (CL) care bundle in order to reduce bloodstream infection (BSI). A CL (or a central venous catheter) is a long, thin, flexible tube used to give medicines, fluids, nutrients, etc., for several weeks or longer. The catheter is often inserted in the arm or chest and threaded through a vein until it reaches a large vein near the heart. The CL bundle comprises five interventions in order to reduce BSI: (1) strict hand hygiene; (2) maximal barrier precautions; (3) use of chlorhexidine skin antisepsis; (4) optimal selection of the catheter site, with avoidance of the femoral vein for central venous access in adult patients; and (5) daily review of the line necessity, with prompt removal of unnecessary lines (Institute for Healthcare Improvement [IHI], 2014).

A Way of Reconciling WAI and WAD

A frequently used approach to improve the safety of everyday clinical work is to rely on compliance with guidelines and policies. Because guidelines and policies obviously represent WAI, the effectiveness of insisting on compliance assumes that WAI and WAD are similar or even identical. This is, however, rarely the case and may indeed not even be achievable in practice. In the ICU study, an RHC perspective immediately drew attention to the fact that a proper implementation of the CL bundle had to acknowledge that varying performance conditions would exist and that it therefore would be necessary to accept that cases of so-called non-compliance would be appropriate in some circumstances. Examples of 'non-compliance' are often found in studies of everyday clinical work, and the inevitability of this has often been reluctantly accepted. Thus Gurses et al. (2008) found examples of non-compliance such as not washing hands before touching the patient and not following maximal barrier precautions, and explained this as a result of various types of system ambiguity. Carthey et al. (2011) adopted a more detached view by pointing out that since humans by nature are adaptable and tend to improvise, some degree of non-compliance would be inevitable. They specifically noted that

> Length, complexity, accessibility, volume, and failure to consult with healthcare professionals who have to follow a policy all reduce compliance with potentially critical results. Staff may break the rules because of their complexity, may follow the wrong policy when there are multiple versions, or be completely unaware of a policy because of its obscure location or inadequate dissemination. (Carthey et al., 2011, p. 3)

From an RHC perspective, the term *non-compliance* is actually misleading, because its use presumes that WAD should correspond fully to WAI, as represented by the guidelines. The term also implies that the difference is attributable to the person (or persons) carrying out the work, although some studies make allowance for 'unintentional non-compliance' (Barber, 2002). Yet due to the inevitable difference between WAI and WAD, absolute compliance is an unrealistic requirement. People do improvise and do adjust their work to the situation, to make it possible under the given circumstances. But such adjustments do not represent non-compliance. On the contrary, they are necessary in order for work to be carried out under varying circumstances.

An RHC perspective requires that efforts are made to identify those conditions that make performance adjustments necessary – or, from a different perspective, that make compliance with bundle requirements difficult or impossible – and determine the types of adjustments that could be useful in those conditions. This in turn means that it is necessary to learn from the majority of cases where everyday clinical work goes well rather than bemoan the few where it does not. Safety can be improved by facilitating the useful and sometimes ingenious performance adjustments that are the foundation for everyday clinical work, but not by pushing the clinical staff even harder to comply with the procedures in the Safety-I tradition. Indeed, investing extra work in checklists and audits in order to increase compliance rates only makes sense if the discrepancy between WAI and WAD is negligible – and that is only rarely the case. RHC instead advocates the development of a holistic framework that can integrate Safety-I and Safety-II perspectives and thereby guide the practical implementation of improvement initiatives such as the CL bundle.

One aim of the ICU study was to emphasise that health care workers should use and adhere to the CL care procedures and make use of performance indicators to increase compliance rates in regular cases. Another aim was to avoid as far as possible that patients were harmed due to the number of local, situational variations. To accomplish the two aims, the study followed the regulations including performance indicators that had been set by the CL bundle programme from the Centre of Disease Control (CDC) and the project management in the hospital. Compliance with the CDC indicators was associated with an incentive reward for the hospital. The study also introduced three additional indicators to measure Safety-II safety and 'non-compliance' interventions. However, collecting the three additional indicators with related interventions was not mandatory because the ICU staff were not obliged to take part in the study. It was nevertheless still expected that the staff would participate in the research.

Which Activities Went Well and Which Did Not?

The results from the first year of the study did not show any reduction in the number of BSI in the ICUs. The BSI infection density was 9.46‰ in 2012 and 10.1‰ in 2013 in the surgical intensive care unit (SICU), and 10.21‰ in 2012 and 9.9‰ in 2013 in the medical intensive care unit (MICU). This may have been due to a strong regulation of the CL bundle because the interest of top management was to achieve high compliance. The study was therefore conducted with support from a limited number of ICU staff, but without management providing the proposed supportive framework. It may also have been due to the unexpected resignation of a couple of ICU nurses in the period. Despite the lack of change on the main outcome measurement, the implementation of the framework gave rise to some noticeable findings in how the front-line services were carried out:

- There was an acceptable compliance rate, which was neither too high nor too low. There will always be some CL bundle steps where compliance is either impossible or impractical.
- More ICU staff participated in the CDC-required activities (including CL bundle training and team resources management training) than in the seminar or round-table discussion held by the research team.
- Discussion of workarounds identified the possible responses to solve the issues of optimal catheter-site selection and the use of 2% chlorhexidine. The use of 2% chlorhexidine for skin preparation to prevent surgical infections was a new requirement by the CDC.

But the use of 2% chlorhexidine was a cost concern because it was more expensive than the original sterilising fluid used by the hospital. The choice of whether to use 2% chlorhexidine or the original fluid was left to the ICU staff. A possible workaround would be to change the insertion site after several tries in subclavian vein into femoral.

• The CDC programme pushed ICU staff to focus on the actual CL insertion and its post-operation care rather than on the judgement of insertion decision for a patient.

The CL care bundle is a recognised standard for reducing BSI, and hospital management usually requires staff to follow the standards and procedures, except for some situations, e.g., the purchase and use of 2% chlorhexidine, since that increases the cost. The ICU study revealed that the proposed strategy worked well for the existing allowable workarounds, although not all of them contribute to increase the organisation's resilience. This indicated that reconciling WAI and WAD is better achieved by encouraging and strengthening existing 'good' practices than by a radical change to existing ways of doing things (as compliance would represent). Activities that need extra staff efforts to balance WAI and WAD, particularly in the cases such as the ICUs where human resources are a critical limitation, would not succeed unless the top management provides full support and staff would like to learn new things.

Major Obstacles

Health care organisations conventionally rely on evidence-based guidelines to prevent hospital-associated infections (HAIs). Such guidelines include well-recognised factors that facilitate institutional oversight of HAIs and nosocomial epidemics as a part of the structure and components of an infection-control programme, such as the CL care bundle. In addition, root cause analysis (RCA) is often used to complement an epidemiologic investigation of infection-related adverse events (Carrico and Ramirez, 2007; Gerberding, 2002; JCAHO, 2000; McKee and Macleod, 2012; Storr et al., 2013). Taken together, the two approaches provide epidemiologic and managerial information for infection prevention and control (Frain et al., 2004). Because of the widespread reliance on such solutions, hospital management as well as staff have become used to and therefore tacitly accept these kinds of procedures or approaches.

The ICU study took a different approach to implement the CL care bundle and thereby improve BSI. Instead of striving for high compliance rates through conventional guidelines, training was provided to introduce the principles of RHC. A system-oriented event analysis model of how a CL care system works was also explained. The experience from attempting this points to some potential obstacles for such a strategy:

- The broad concept of RHC is not easily understood or accepted by health care professionals, regardless of their level in the organisation.

  The current broad concept of RHC comprises theories, a number of practical issues in health care systems and suggestions for improvements that can be made. But in addition to these there is a need of an overall 'picture' that clearly explains how the RHC can realistically be embedded in the health care systems. Without such an overall picture, people at different roles in a hospital will naturally base their interpretation of what RHC is on a partial and subjective understanding. Experience shows that this partial understanding can lead to one or more of the following concerns or questions. One concern is that resilience engineering is understood as an engineering discipline, which therefore might be difficult for health care professionals to adopt or accept. Another issue that people are uncertain about is whether RHC is 'merely' a philosophy or whether it offers concrete solutions – preferably ready-made and immediately applicable. Yet another is that while many people agree with the Safety-II perspective, they are worried that it may not be possible to analyse all the many things that go right – even though that is not the intention of Safety-II. Others argue that since management-by-exception – which means focusing on situations that deviate from the norm in one way or the other, such as cases of non-compliance – seems to work, there is no need to adopt new methods especially when resources are limited. And finally, people sometimes posit that they already have implemented workarounds for health care workers under certain conditions, that they have meetings to discuss the near misses that actually went right, hence that they already are resilient.

  These obstacles can be countered by targeted communication about RHC and by practical introductory courses in the methods and principles of RHC. Such communication and training should be designed so that it is tailored to different levels of staff and managers in hospitals. An important part of such training could be the demonstration of successful cases of implementing RHC.

- RHC generally argues that the traditional event analyses based on linear cause–effect relations, such as RCA, are not effective and that this is one reason for the slow improvement of patient safety (Cook, Woods and Miller, 1998; Hollnagel, 2011a). This argument can be a serious obstacle to the introduction and implementation of RHC and it is often met with the counterargument that such traditional analyses are not ineffective. The background is that health care organisations for many years have been compelled by the requirements of hospital accreditation programmes to use the traditional approaches. Most health care staff have been trained in these methods and are therefore accustomed to them. It is therefore not surprising that they are reluctant to accept the RHC argument that patient safety is an emergent rather than resultant property of health care systems. The limitation of traditional approaches means that an RCA at best will reveal just one or a few of the components that affect the system's performance. The alternative is event analyses that are based on a different and unfamiliar approach that initially will be more cumbersome.

In order to counter these arguments and effectively promote RHC or reconcile WAI and WAD, it is important to point out that the traditional and the new approaches – and models – should be seen as complementary rather than contradictory. The new approaches are not intended as a wholesale replacement of what is already there, but as an addition for cases where the established methods come up short. RHC thus represents an evolution rather than a revolution.

In order to communicate this message effectively throughout an organisation, both lower-level learning (single-loop) and higher-level (double-loop) learning are needed (Fiol and Lyles, 1985; Argyris and Schön, 1996). Lower-level learning tends to take place in circumstances that are well understood and which are controlled by management through partial and local changes with short-term impact. Whenever something goes wrong, an initial response is to look for another strategy that will address the situation and resolve the problems. Higher-level learning occurs when an organisation tries to adjust overall rules and norms with long-term effects for the whole organisation. This takes place by subjecting the conditions of the event to critical scrutiny, possibly leading to a shift in the way in which strategies and consequences are framed. RHC argues that methods for event analysis should not only strive to find concrete solutions, but also recognise the organisation's genuine needs for learning over a longer time frame (Tamuz et al., 2011; Duncan, 1974).

- Difficulties are encountered in determining feasible measurements for RHC. In the ICU study it was difficult to find appropriate indicators without at the same time creating extra workload for the ICU staff. People often find it difficult to propose indicators for acceptable or successful outcomes – the things that go right. While it may be difficult to develop meaningful indicators that support the purposes of Safety-II, it is actually no more difficult than it is for Safety-I. The difference is that for Safety-I we have become used to doing things in a certain way and have therefore stopped thinking about it. Most of the safety indicators that are used, for instance in the Organisation for Economic Cooperation and Development (OECD) guidelines, count how many times a specific activity goes wrong or how often adverse outcomes occur. (The hospital standardized mortality ratio [HSMR] is a case in point.) There are several other existing indicators used by ICUs, hospitals, regulators and even political institutions. These indicators are used because they are convenient and easy to use, rather than because they are meaningful. Because there is no corresponding tradition for Safety-II, we are faced with the problem of proposing indicators or measurements that are meaningful rather than convenient. And this can, not surprisingly, be rather difficult. To create new indicators and find ways for the ICU staff to collect them was therefore not achieved in the ICU study. Because the starting point was that the collection of CDC-required indicators was requested by the top management of the hospital, developing new indicators was an uphill battle.

The simple way to overcome this obstacle is to look for measures of how often something goes well, rather than how often it fails. Although this is not often done routinely, it is not really difficult to accomplish. Another solution is to consider the four cornerstones of resilience (Hollnagel, 2009b), i.e., the abilities to respond, monitor, learn and anticipate. This would be a step away from a simple outcome measure (the number of BSIs) and towards a process measure, or at least a measure of the abilities that are the basis for resilient performance. A process measure will directly support the management of a planned change both by making it possible to see discrepancies at an earlier stage and by making it much easier to determine the necessary corrections.

- Leadership involvement is necessary. In order to implement RHC in an organisation, management support is needed for training, data collection as well as communications between departments. If such support is not given, the implementation of RHC is very likely to fail. But management support can be difficult to get because the time and availability of top-tier managers are usually very limited. People who work in that capacity rarely have enough time to listen to presentations and explanations about RHC, let alone participate in RHC training courses.

To overcome this obstacle, two things are needed. First, brief but succinct presentations of RHC are developed – the so-called 'elevator pitches'. Second, the presentations can be illustrated and supported by 'success stories' that document the practical benefits in real-life cases. Such 'success stories' will highlight the principal and practical differences between how everyday clinical work is performed and improved using traditional approaches and how RHC can be planned and developed in an effective manner. Examples of this nature may be useful to create their vision and raise their involvement.

Suggestion and Conclusion

Because we are still in the early stages of RHC, there are a few published empirical cases to use as examples and to support the theoretical arguments. Continuous training of resilience engineering and developing a prospective strategic implementation plan in organisations are necessary. Therefore, an explicit conceptual model approaching RHC is particularly important.

To help overcome the obstacles found in the study, a holistic practical conceptual model of RHC should be developed. Health care practitioners and researchers can use this in the design of RHC training and studies, as well as to facilitate a 'how to' map to implement RHC in patient safety management.

A pragmatic suggestion for how to reconcile WAI and WAD would be to start by integrating existing methods and traditional cause–effect analysis approaches with RHC concepts into a suitable approach that fits the organisational status. This should be followed by gradually modifying the combined approach as the maturity of RHC in the hospital grows. If we characterise the current patient safety management as a corrective approach, then reconciling WAI and WAD will be a facilitating approach, which can facilitate the combination of methods and procedures that provides system resilience. The process of going from a corrective approach to a facilitating approach will need to take place in stages, and be carefully monitored, for instance using the resilience analysis grid.

Part II

Applications

Jeffrey Braithwaite, Robert L. Wears and Erik Hollnagel

Application (n). 1.a. The action of bringing something to bear … with practical results; the action of causing something to affect a person or thing … 3.a. … putting something to a use or purpose. … (Oxford English Dictionary [OED]: http://www.oed.com/view/Entry/9705?redi rectedFrom=Application#eid.)

Having established some baseline points of view and teased out a selection of core concepts concerning the challenges of reconciling work-as-imagined (WAI) with work-as-done (WAD) (and vice versa, of course), we move to a set of chapters that focus on *Applications*. The OED definition underlines a core point that each chapter writer grapples with here: how to put their ideas, theories or empirical findings about reconciling WAI and WAD to use for real-world purposes.

The six chapters that follow illustrate the diversity of what it means to apply ideas about this reconciliation in practice. The most obvious thing readers will see is that these exemplar applications range widely: looking at identifying patients requiring blood transfusions or medication, for instance, as Nakajima, Masuda and Nakajima do in Chapter 6. Stringent compliance with WAI, Nakajima and colleagues observe, is not a solution. Even with an activity as apparently straightforward as labelling patients, according to Nakajima et al., it is better to facilitate everyday clinical work and leave practitioners with some degrees of freedom to adjust to circumstances than try to compel people on the front-line to do things under strict prescription.

Another way forward is to consider how the world looks from the perspective of multiple players (in the case of Chapter 7, in a large hospital in Townsville, Queensland, Australia). Johnson and Lane tease out how clinicians, middle managers, senior managers, bureaucrats, health ministers and of course patients and families, see things differently from each other. Johnson and Lane provide a model to help bring these views together: the Ten Cs model. The Ten Cs (cohesion, capture, cognition, communication, culture, clear ownership, constraints, challenge, competence and compliance) are akin to traits in humans: features, for Johnson and Lane, of the system they seek to model.

Discipline-specific case studies can illustrate various aspects of resilience that we can delve into and absorb. It is abundantly clear from chapters in the two previous books about resilient health care (RHC) that the breadth and depth of emergency care, and its characteristics as a time-critical, pressurised and busy environment, represent an excellent source of understanding resilience as it unfolds in complex settings. This is demonstrated once again by the work of Braithwaite, Clay-Williams, Hunte and Wears in Chapter 8. Emergency care is a domain that requires considerable front-line flexibility. Clinicians apply customised treatments in unique configurations. They are bricoleurs, stitching things together in their domain of activity to traverse the WAI–WAD differences.

By virtue of its internal logic, incident reporting is a Safety-I tool focused on special occurrences – things going wrong – with the goal of reducing or eliminating future harmful events. But even though incident systems reporting harmful events have been the focus of people in health systems to date, there is nothing to stop us from developing reporting systems to register ordinary, everyday events, and learning from these. Sujan, Pozzi and Valbonesi in Chapter 9 examine just such a system. It records small-scale, frequent hassles that everyone experiences in their daily lives. Running an incident system recording special events alongside an "everyday events reporting system" could help overcome the WAI–WAD differences by providing an understanding of real-time, frequently occurring clinical activities and not only those that occasionally cause harm.

As they go about their tasks, both clinicians and managers develop repertoires – a store of regularly used behavioural algorithms. Mostly, according to Cook and Ekstedt in Chapter 10, resilience is enacted by practitioners expressing it in their work activities rather than resilience being enabled by people at the blunt end trying to influence ground-level behaviours. Repertoires develop after prolonged exposure, such as through apprenticeship models that proliferate in medicine, or by people responding to a sufficient range of disturbances and learning from them.

Of course, people responsible for preparing, organising and managing work at the blunt end can, and do, develop repertoires. They can refine their repertoires by receiving feedback on their efforts in suggesting, prescribing or mandating solutions, and observing and acting on their effects on the front line. This will take longer of course, because people at the blunt end do not receive timely feedback – there are typically significant delays. And in addition to that, the feedback is nearly always impoverished and reduced to a few simple categories. As they conclude their analysis of repertoires at work in health systems, Cook and Ekstedt ask a penetrating question. To what extent can resilience actually be engineered by external agents who have to rely on descriptions of WAI? They are themselves sceptical, but others including writers of chapters in this book and previous RHC volumes think blunt-end participants can influence and shape coalface clinical care in ways that enhance front-line expressions of resilience.

This brings us to the nature of power in health systems. Power can be defined in many ways, but for our purposes here it is capacity for action. Hunte and Wears argue in Chapter 11 that sufficient power is needed in all sorts of circumstances in which work is to be accomplished – in self-organising, caring for patients, making trade-offs, selecting among alternative priorities and then enacting them, accommodating to circumstances (e.g., demands) and improvising. Although clinicians sometimes look very powerful (e.g., to patients), the reality is that they are always limited in their capacities, but at the same time also enabled to do things by virtue of their agency (that is to say, their roles, their knowledge, their expertise and their relative autonomy).

Although it is true that front-line clinicians have power by virtue of being relatively self-determining professionals, and also that society through its legal structures confers power to act on professionals in all sorts of ways, ultimately including via direct legislation, WAI proponents in policy and management circles provide a deal of the structures and resources by which clinical agency is possible. Hunte and Wears draw our attention to a broad distinction, not wholly complete, but with a large grain of truth, that clinicians have power *to* (do things) rather than power *over* (other senior stakeholders). Accordingly, for Hunte and Wears, generative partnerships with interdependence at the core are needed if resilience is to flourish. A partnership model is therefore a prerequisite for bringing together blunt- and sharp-end interests. This in turn implies that we will need a more keen appreciation of how power works if we are to strengthen health care: both WAI and WAD stakeholders must also be willing to share and relinquish power.

6

Exploring Ways to Capture and Facilitate Work-as-Done That Interact with Health Information Technology

Kazue Nakajima, Shinichi Masuda and Shin Nakajima

CONTENTS

Introduction

Work-as-imagined (WAI) is essential when we plan or manage work, and should represent our best understanding of what work will be like. Work-as-done (WAD) represents the reality of everyday work, of what actually is done in specific situations. Because the differences between WAI and WAD should not be too large, the reconciliation of WAI and WAD is key to improving the performance of organisations as well as of individual clinicians. For example, surgeons can revise their own WAI (i.e., surgical planning) by learning from their own experience of WAD. Such improvement in WAI can lead to successful WAD (i.e., operations) and allow individuals and teams to anticipate and prepare for possible events.

Resilience engineering explores ways to reconcile WAI and WAD in complex systems that involve significant interactions between humans and non-human agents, but may also be applied to single individuals or teams (Hollnagel et al., 2013). Health care systems are characterised as not only complex, but also adaptive, with each system comprising multiple smaller systems that are organised hierarchically and/or heterarchically. The dynamic interactions among agents on different levels of many systems, where people often improvise and behave adaptively based on what they have learned, sometimes lead to the emergence of unexpected outcomes (Braithwaite et al., 2013a; Robson, 2015). In trying to take a Safety-II approach to patient safety in such complex adaptive systems, we are first faced with the difficult question of how we can capture WAD in everyday clinical work. This is a potentially overwhelming task due to its complexity and size. Hollnagel (2015b) has suggested possible data sources that could be utilised for this, including event reports, observations, interviews, questionnaires and simulation. He also proposed two principles for analysing complexity: breadth-before-depth and frequency-rather-than-severity. In this chapter, we describe a few methods for capturing everyday clinical work, particularly looking at the use of information technology in patient identification in blood transfusion and medication administration. We then identify gaps between WAI and WAD in the procedures and analyse the reasons underlying performance adjustments. Finally, we propose a potential strategy for facilitating everyday clinical work by examining control signals from technologies to clinicians using an instantiation of a functional resonance analysis method (FRAM) model.

An Exploratory Journey to Capture the Reality of Everyday Clinical Work

Starting from the Investigation of a Specific Event

In the previous book, *The Resilience of Everyday Clinical Work*, Nakajima (2015) described two cases involving medical adverse events. In each of these, a critically ill patient in the hospital emergency department (ED) was given a blood transfusion with a blood component of the wrong type that had been prepared for a different patient. This happened despite the implementation of barcode technology to verify patient identification. In case 1 the personal computer (PC) terminal froze while the patient's condition seriously worsened, and the nurse bypassed the barcode reader and administered the blood component to the patient without manually performing the necessary patient identity checks. In case 2 the nurse manually entered many-digit numbers unique to each blood component into the electronic health record (EHR) systems after the administration of blood products to the patient, although this was supposed to be done using barcode technology prior to administration. (Case 2 happened at our hospital; case 1 happened at another hospital.) Conventional investigations identified how and why the clinicians failed to identify the patients in these specific cases. In addition, the investigations successfully obtained information about the conditions or situations that compelled clinicians to make adjustments in performing their functions related to specific tasks in everyday clinical work. For example, in both cases, various hospital policies for blood transfusion, nursing procedures and hospital information systems existed independently, but did not comprehensively explain how to check the match between a patient and the blood components intended to be administered in a specific context (such as the ED) in using health information technology. In other words, the hospital policies (i.e., WAI) did not reflect the actual practice (i.e., WAD), which resulted in a wide range of improvisations and adjustments in clinicians' performance.

Trying *In Situ* Simulation in a Specific Place

Following case 2 an *in situ* simulation of massive blood transfusion for a mannequin patient with multiple traumas in the procedure room of the ED was carried out in order to develop an operational procedure for blood transfusion with barcode technology that would be feasible in the ED. We attached importance to interactions between clinicians and the EHR systems in the simulation. Therefore, we prepared dummy patient information, to be registered in the hospital EHR systems as an admitted patient, as well as dummy blood components whose information would be used afterwards for computerised verification.

The *in situ* simulation illuminated several gaps between WAI and WAD in patient identification using barcode technology. Clinicians were not knowledgeable about which of several barcodes on the blood component bag should be scanned. In addition, the soft and wet surfaces of the bags were not easy to scan with the barcode reader. Use of hand-held computers with the barcode reader seemed unsuitable for verification in the crowded intensive care environment, because it would require them to be operated in close proximity to the patient under sterile conditions. Instead, a wireless barcode reader with a movable computer terminal was preferred. Massive transfusions also required very quick verification and readiness to rapidly administer blood components through a fast-flow infusion pump. The *in situ* simulation identified factors that may influence clinicians' function in using barcode technology for patient identification: their knowledge of the operation of barcode readers, the performance or usability of the device and the suitability of the device for the workflow and environment.

Expanding *In Situ* Task-Limited Simulations in Other Wards

After the *in situ* simulation in the ED, we conducted a similar *in situ* simulation in non-emergency inpatient wards in our hospital (Dan et al., 2012).

We limited the observation to patient identification processes involving administration of blood components; by contrast, the previous simulation in the ED included the whole process, from the patient's arrival to the blood component administration. The patient safety committee members visited each of 24 inpatient wards and asked one clinician (a physician or a nurse) to administer two units of red cell concentrates (RCCs) to a patient as he or she normally did. A committee member in the round wore a wristband with the information of a dummy inpatient. The seven real (but empty) bags of RCC included five "contaminated" bags intended for a different patient. The hospital policy regarding blood transfusion recommended that clinicians must visually identify the patient, visually check the patient's name and the blood type of the intended blood components, and verify the match between the patient and blood component using barcode technology (either a hand-held computer or a barcode reader connected to a PC terminal) before administration.

The observation found that 6 out of 24 clinicians verified the patient identification with hand-held computers while eight used barcode readers with PC terminals. Another 8 clinicians directly input multi-digit numbers from the blood component bags into the PC terminal manually; this was also an acceptable procedure in situations when the barcode reader failed to scan the barcode of blood component bags, when it did not function due to radio interference or when it was not available. The rest did not use the EHR systems before administration, but instead input the data after the administration. An interview with a clinician who used the EHR systems after administration reported that the wording of the blood transfusion policy, "Input the data of the administration of blood components into the EHR systems" was vague about the purpose and relative timing of these actions, e.g., whether the input should be performed immediately before administration of blood components (for patient safety) or sometime after administration (for documentation). In other words, the health information systems related to one action (reading a barcode) involved several different functions, including patient identification and documentation; these systems were efficient, but not designed with sufficient sophistication to indicate the meaning and timing of the action. Thus, the simulation captured WAD, which differed from WAI because of unclear language in the procedure manual, poor cognitive design of the specification of the use of the barcode technology or mechanical problems related to the input/output devices of the EHR systems.

Questionnaire Survey on Patient Identification in Medication Administration

In our hospital, intravenous medication requires a similar procedure for patient identification. To understand adjustments related to this process, we surveyed nurses to find out how they use information technology when administering intravenous medication, using the anonymous questionnaire function of the hospital e-learning system (Shimai et al., 2011). During 2 weeks, 668 nurses answered the questions. According to hospital patient safety policy, the recommended way to verify patient identification in medication administration was to use either hand-held computers or barcode readers connected to PC terminals immediately before administration; direct manual inputs to PC terminals were also permitted.

Of 552 nurses who had used the EHR systems for medication-related tasks, 241 reported that they made direct inputs into the EHR systems without using devices with barcode readers, whereas 204 used hand-held computers and 57 used barcode readers with PC terminals. Of the nurses who responded, approximately 50% reported that they made direct inputs into the EHR systems without using devices with barcode readers, whereas 40% used hand-held computers and 10% used barcode readers with PC terminals. Many free comments reported conditions and situations in which WAD at the sharp end was different from WAI described in the medication policy. Clinicians' major frustrations in using hand-held computers were their slow speed of operation and verification, difficulty in reading the barcodes, insufficient numbers of the devices and short battery life. Their comments also mentioned the unwieldiness of the devices due to their relatively large size, and difficulties in seeing the display due to the small size of the letters. With respect to the use of barcode readers with PC terminals, nurses felt that it was inconvenient to carry movable PC terminals to patients' bedsides (in rooms containing up to four patients) solely for the purpose of using barcode readers at every intravenous medication. In addition, they mentioned scanning or battery problems, as also experienced with hand-held computers. They also expressed hesitation on the grounds that the proximal operation of barcode readers and the sound of verification might bother sleeping patients, particularly at night.

As a result of their adjustments to these conditions, half of the nurses directly input the medication information without using barcode technology, mainly at nursing stations where PC terminals were located. Among those who input the data into PC terminals without barcode technology, approximately 30% input the data before intravenous medication was started, and the rest did it afterwards. This survey revealed that although barcode technology had been implemented to enhance patient safety, including the accuracy of patient identification before medication and blood transfusion, many clinicians confirmed patient identification in the conventional way (i.e., visually) and input the data into the EHR systems after administration for documentation; they behaved in this manner due to technological, environmental, workflow and other considerations.

Importance of Control Signals from Technology to Clinicians

To understand the interactions between clinicians and health information technology, we instantiated a FRAM model of the patient identification process for administering blood components. Each hexagon represented a single person or single piece of equipment, so that we could readily illustrate that each entity has multiple functions within a certain period of time. We also labelled each hexagon to readily identify each specific person or piece of equipment. The health information system was treated as an independent, non-human agent that interacts with human agents, rather than as a tool used by human agents. Control signals were carefully distinguished from input signals. Like traffic lights, control signals – technologies, paper-based forms or verbal communication – must be accurate so that people can make appropriate judgements and behave accordingly.

Figure 6.1 shows a small system involved in the administration of blood components. The system is composed of a clinician and an input/output device (a hand-held computer or a barcode reader with a PC terminal) of the EHR system. A clinician has several functions:

- To check patient identity by asking the patient to give his or her name
- To visually match the patient to the blood components
- To operate the device and scan the barcode of the blood components

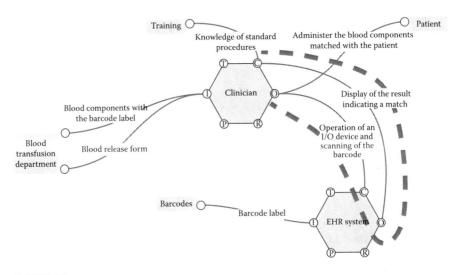

FIGURE 6.1
A FRAM model showing the patient identification process for blood transfusion using barcode technology.

- To verify the match between the patient and blood components using a barcode reader
- If barcode technology is not available, to verify the match between a patient and blood components using a blood release form as reference information
- To administer the blood components

The EHR systems also have several functions:

- To read barcodes of blood components
- To display the result indicating a match between the patient and intended blood components
- To record the administration

The lines connecting the two function symbols representing the clinician and the EHR systems show that the clinician controls the EHR systems by operating an input/output device to scan the barcode, while the output from the EHR, the display of the verification result, controls the clinician in verifying the match between the patient and blood components for the following blood transfusion. The dotted thick line highlights the feedback loop. A similar model could be applied to administration of medication.

The FRAM model of the local system indicates that outputs from the technological systems influence the clinician's functions through the interaction loop between them. As long as high-quality control signals from the technologies reach clinicians in a timely manner, clinicians can execute their functions appropriately. If this fails to happen, however, performance variability is inevitable. In clinical work in a time-critical environment, such as the ED, even a small problem in the technology imposes additional functions on the clinician, who must diagnose and solve the problem in a limited time, potentially exceeding their performance capacity. Once clinicians experience a situation in which control signals are not delivered in an efficient manner, they adjust their behaviour by adopting an alternative method in the future, e.g., avoiding use of barcode technology regardless of the recommended policy.

Possible Methods for a "Breadth-before-Depth" Approach

Resilience engineering must identify adjustments and understand why they are made in everyday clinical work – the "breadth-before-depth" principle. Learning should be based on frequent events rather than severe but rare occurrences – the "frequency-rather-than-severity" principle (Hollnagel, 2014b). The question is how this can be achieved. Our exploratory journey suggested that information about adverse events or incidents can be a good starting point to look for performance adjustments in repetitive or regular clinical activities, although conventional investigations are limited by confirmation bias (i.e., "what you look for is what you find", cf., Lundberg et al., 2009). Triggered by the two cases described above, as well as other related incidents, we developed an interest in processes involving patient identification using information technology, initially in the context of blood component administration and later in the context of medications; these are the most frequently performed procedures in everyday work in an acute care setting.

Once the target of the analysis is decided, WAD can be captured through interviews or observations of *in situ* simulations or ward rounds conducted with an open-minded attitude. In collecting data about performance adjustments and analysing the complexity of work, sociotechnical aspects should be considered. We believe that *in situ* simulations (even if task limited and brief) provide an opportunity to observe adjustments in a specific clinical context and understand the situations underlying performance variability. Questionnaire surveys, when they are carefully designed to obtain both quantitative and qualitative information, also seem helpful in illuminating a variety of reasons for WAD. Our study suggested that a combination of several established methods, including case analyses, observations of and interviews with clinicians in *in situ* simulations, and questionnaire surveys represent a valid approach to comprehending the complexity of everyday clinical work.

Barcode technology linked with EHR systems has been introduced to enhance error detection in patient identification. In WAI, errors may be detected more reliably with technology than without. Our observations and interviews during the *in situ* simulation and subsequent questionnaire survey revealed clinicians' adaptive behaviour in the use of information technology. Barcode technology, implemented to enhance patient safety, was not utilised as imagined for various reasons including unavailability, unfulfilled performance expectations and unwieldiness of the input/output devices. Instead, the data were often input into the EHR systems for the record, whereas the final verification of the correct patient and intended treatment remained dependent upon humans' checks. Such situational adjustments, which arise due to both technological and non-technological factors, result in successful accomplishment of everyday clinical work in nearly all cases, but failures on rare occasions. The risks arising at the sociotechnical intersection between patient safety and health information technology (Meeks et al., 2014) are topics that should be carefully studied from a Safety-II perspective.

Resilience engineering aims to increase the numbers of things that go right, rather than just to avoid recurrence of similar adverse events in the future. The examples provided here show that it is not a solution to compel clinicians to use barcode technology for patient identification and monitor their compliance. Instead, we must facilitate everyday clinical work. One potential strategy is to improve the performance and reliability of input/output devices, making them more suitable for the workplace context, so that clinicians can receive high-quality control signals from health information technologies in a speedy and efficient manner. It also would be useful to design interfaces of health information technology to clarify the intended purpose of specific operations of computer devices (e.g., for either safety or documentation) and to navigate clinicians' workflow (e.g., before or after administration). Furthermore, it is necessary to develop a single and integrated procedure manual for a specific task using EHR systems, rather than having multiple and independent policies relate to the same task with little consideration for interactions with technology.

Conclusion

In the traditional normative approach to patient safety, increasing compliance and strengthening education are often emphasised based on the assumption of poor clinical competence of individuals, as if they were operating in a stand-alone environment. Resilience engineering needs to explore ways to elucidate the reality of WAD, understand reasons for adjustments and facilitate everyday clinical work. Modern health care is a complex adaptive system in which interactions with technology are inevitable. According to the breadth-before-depth and frequency-rather-than-severity principles of the analysis of everyday work, control signals from technologies to clinicians represent a major target for systemic improvement in the era of health information technology.

Acknowledgement

The authors would like to express great appreciation to Aoi Uema, RN, MS, for her professional support on health informatics.

7

Resilience Work-as-Done in Everyday Clinical Work

Andrew Johnson and Paul Lane

CONTENTS

Introduction

Health care professionals and therefore the health care system, 'do' resilience all the time (Nemeth et al., 2008; Wears and Vincent, 2013). The collective individual actions come together to create resilience within the system. It is a challenge to define what individual resilient traits actually look like. This chapter aims to first offer a different perspective on everyday clinical work (ECW), suggesting that ECW occurs at all levels within the health care environment, not just on the clinical 'front-line', then characterise examples of resilient qualities of the individual. We will propose a taxonomy for resilient qualities in a model showing how these qualities allow individuals to compensate for system disturbances and for gaps and weaknesses in their environment.

What Is Everyday Clinical Work?

The standard logic would suggest that ECW is what 'front-line' clinicians do every day in their care of patients, and it certainly is – in part at least. ECW is, however, much more than that. We suggest that ECW is practised at every level within the health care system from Minister for Health to clinicians providing direct patient care. We sometimes like to think about the population health focus of the policy makers as the blunt end, while individual patient care represents the sharp end of health care delivery (Hollnagel et al., 2013). This thinking is akin to a cognitive bias. The labelling of individuals may act as a hindrance to collaboration and cohesion.

The Sharp Reality of Life at the Blunt End

Health care is delivered along a continuum from an individual patient focus to a population focus, the objectives, time frames and mechanisms for both ends of this spectrum are very different, and the reality is that both ends are involved in delivery. In between sit the support players who are challenged to deliver at both ends of the spectrum. It could be said that there are multiple sharp ends and that "everybody's blunt end is someone else's sharp end". Therefore, it is important when using these terms to understand that the distinctions are relative rather than absolute.

In a hierarchical model, those who work at the highest echelons of the politico-bureaucratic enterprise are undoubtedly working as hard or harder than those at the 'front-line'; however, their work is different and the function is no less important. At this theoretical 'blunt end', workers commonly work long hours, responding to demands from political and bureaucratic masters for clear and accurate information, upon which they will base policy decisions. At the clinician end, the stakes are one life and one family within moments. In contrast, at the policy end, decisions affect entire populations over years (Figure 7.1).

Such a model invites us to recognise that Hollnagel's premise of information delay, decay and half-life is bi-directional. That is, the delay in transfer of relevant information 'up' the 'chain of command', with consequent devaluing of the information and therefore lack of relevance to current circumstance is matched by delay in transfer of policy direction down the chain. These time delays have an important influence on decision-making by the 'blunt end' that is poorly understood. A delay in decision-making is easily perceived as a failure to listen, a failure to care and a devaluing of the relationship. Yet this is usually not the case. Indeed, when clinicians make quick decisions in seconds, it can be perceived as unconsidered and of poor judgement by those further along the continuum.

We agree that health care is a complex adaptive system (CAS) (Clay-Williams and Braithwaite, 2013); therefore, various microsystems and macrosystems

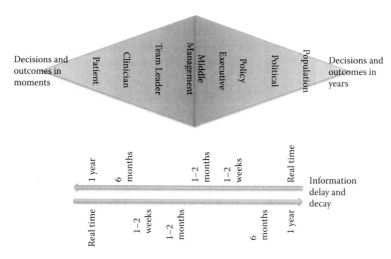

FIGURE 7.1
Health care delivery continuum, information delay and decay, decision time frame.

(Barach and Johnson, 2006) exist at the same time but evolve over different time scales. In reality a quick decision in health care can take seconds at one end of the continuum and months at the other end.

Further, this model gives weight to the emergent problem solving often seen in health care. This is particularly relevant for those at the clinical end who must deal with ever-changing circumstances with an acknowledged delay in information flow from above.

And finally we have the 'jugglers', those in the middle of the continuum. These support people are destined to serve two masters and they face a virtually impossible task. They have to translate, interpret, shield and deflect in an effort to hold things together, to maintain the relationship. Their role, while seldom recognised, is vitally important.

To have effective relationships, we all have to understand the important and challenging work of the individuals in other parts of our system. We need to avoid pejorative characterisations that suggest that one end of the system is less important than the other. As we will shortly describe, the development of mutual respect through understanding is vitally important for resilient health care.

Let's All Take a Walk on the Wild Side

- Take a walk in the shoes of the clinicians navigating the system, who are asked to follow so many rules that they wake up in screaming nightmares about how many they will break today.

- Take a walk in the shoes of the 'jugglers' of middle management who struggle every day to hold together the relationships with the clinicians they support and the bosses they answer to. Loss of trust in either direction often proves to be disastrous.

- Take a walk in the shoes of senior managers who have performance targets for just about everything that can be measured and some things that cannot.

- Take a walk in the shoes of the senior bureaucrats who can be sacked for any or no reason, who work ridiculous hours trying to make sense of the world so that they can offer a policy response to it.

- Take a walk in the shoes of the Health Minister who gets advice from everyone, most of it contradictory, who has to work out how to make the population as healthy as possible, or at least seem to.

- And most importantly, take a walk in the shoes of the patients and families who depend on us to make sense, provide consistent messages and offer a map and a compass to guide them in the decisions that ultimately (within bounds) are theirs to make.

These different perspectives can create a misalignment of interests and the potential for conflict.

Work-as-Done: The Resilient Way

Individual examples abound of resilient qualities; the following vignettes may ring true, exploring resilience from the perspective of three of our players.

Senior Manager with Performance Targets

Performance targets have become a fact of life for the hospital manager. Patient care has become a precarious balance between efficiency-driven processes and delivery of best practice, safe clinical care. Today, at your weekly senior executive meeting, the chief executive officer (CEO) delivers the news; the hospital is failing emergency department access targets. Urgent action is needed. This is your problem, you own it. You know what to do; you bring together key stakeholders to ensure clear communication of the issues and a consistent message. The politics around this matter is obvious, these are highly visible targets, scorecards are produced. But the hospital is struggling, activity is high and you know there has been an increase in medical emergency team calls to the wards that accept these patients from the emergency department (ED). The balance of safety and efficiency is clearly on show, Hollnagel's efficiency-thoroughness trade-off (ETTO) principle in practice (Hollnagel, 2009a).

The team have to stop and think. This is a complex problem, it requires analysis. The measures normally taken have not worked. The appropriate service directors are present; these are key to affecting change in clinical performance.

While the ED information is transparent, you need to know what is happening elsewhere in the hospital, what difficulties are being faced by other departments that may be affecting patient flow. Some staffing issues are identified which can be mitigated. You all discuss worst-case scenarios for your respective departments. It is clear that putting patient safety at risk is not an option. This is a complex situation.

Finally a plan is in place and this is presented to the CEO. It is a difficult conversation but it needs to happen. Afterwards you feel fatigued, it has been a hectic day, but you are relieved this urgent situation has been dealt with.

The Junior Doctor Navigating the System

But the ED remains chaotic. You are a junior ED doctor caring for an unwell elderly patient. Your shift finishes in 30 minutes and you have concerns about the disposition of your patient. Your patient has severe abdominal pain and, try as you might to sort through the problem, the surgical and medical teams continue to argue about whom the patient will be admitted under. The computed tomography (CT) scan report is yet to be completed. Further, the patient's heart rate has risen despite fluid resuscitation. The other vital signs are fine but something just does not feel right. You are worried that your patient is deteriorating. You have tried to raise your concerns with the ED senior doctor but he appears dismissive and is more concerned about getting the patient admitted to the ward in the allotted time. Finally, you decide that enough is enough and you call a rapid assessment team review, a response to the deteriorating patient in the ED. The team arrive and after the ritual cross-examination they begin to assist you. Another bag of IV fluid is given. The CT report finally appears, the patient has a perforated bowel and a diverticular abscess. The surgeon is called and the patient is prepared for the operating theatre (OT).

The Juggling Nurse Manager

On the surgical ward, you are the nurse unit manager awaiting the arrival of an elderly patient after an urgent laparotomy. Your ward is at capacity and two nurses for the next shift have called in sick. You know that three other elective surgical cases are also due back in the ward soon. You have escalated your concerns about patient safety to the executive and are making plans to fill the vacant shift positions. Thankfully, two of the surgical consultants are doing rounds. You have known both of them for many years and have respectfully asked both to consider if any discharges can be made to which they both give positive responses. You reflect upon how lucky you are as some doctors are not as cooperative. Your junior nurses are dealing with a delirious patient and you realize the night team will need extra non-clinical staff to manage the situation safely. You contact support services and plans are made. Next, one of your nurses approaches you to check two units of packed red cells for transfusion. Despite all that is happening around you, you know this is a key event in patient safety, for some rules cannot be broken. The blood is checked and cross-checked, then you move on. Issues keep coming forward that require your attention in relentless fashion. The workload and economic forces continue to exert pressure on your ward yet your staff work in many ways to keep it safe and continue to care for the patients.

These stories accurately reflect ECW. Resilience is displayed but how do we describe it; what are the traits that have led to safe care despite the complexity of the system?

A Taxonomy for Resilient Behaviours: The Ten Cs

We propose the Ten Cs model as a way of understanding resilience in everyday clinical work. The Ten Cs is a multidimensional, interconnected model of health care, where each agent within a system exerts qualities that can influence other agents. Each element (agent or quality) has a consequence on the others and each may compensate for the others.

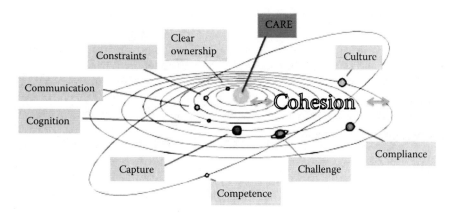

FIGURE 7.2
Cohesion as the gravitational force in the Ten Cs model for resilient health care.

It is a model of a Safety-II (Hollnagel, 2013) system applicable to health care. It describes resilient traits so that we can understand them, teach them and implement them in our practice, translating resilience into work-as-done (WAD).

The Ten Cs are cohesion, capture, cognition, communication, culture, clear ownership, constraints, challenge, competence and compliance (Figure 7.2). As we explain the Ten Cs in more detail, we ask the reader to reflect upon the vignettes above, as well as perhaps examples from their own work in health care, and to use the Ten Cs to identify resilient behaviour.

The Ten Cs describe the qualities available for resilience in each agent of the system (as in the vignettes above). The model proposes the capacity of the qualities of some agents to compensate for deficits in others in times of stress to the system.

For example, in a junior clinical team, gaps in competence may be compensated by strength in compliance with protocol, clear ownership, communication and team cohesion, which in turn develops competence over time. Alternatively a senior team with abundant individual competence may come to grief with a lack of cohesion, poor communication and a culture of fear.

Recognising that it is rare to find perfection, the model suggests that we can cope with weaknesses in some qualities, provided we are aware and ensure that safety is maintained by compensating with other qualities.

The behaviour of the system is determined by the qualities (the Ten Cs) demonstrated by the agents (people, infrastructure) that make up the system. Through awareness of the qualities and the degree to which they are present, we can promote system safety by emphasising those qualities we can access to compensate for those qualities that we cannot access. Through identification of the qualities and the gaps, the learning system will adapt and strengthen.

That is resilience.

The Ten Cs model has emerged in an iterative process of analysis of success and failures of everyday clinical work in a regional tertiary health service over more than 10 years. In part, it has evolved from considerations of the patient safety committee, which brings together key clinical leaders from various disciplines across the health service in a regular meeting that analyses our most significant patient safety issues. The qualities have been proposed, debated, confirmed or deleted, amalgamated or separated as our understanding of their impact has developed.

The qualities vary in terms of their relative importance and applicability in specific circumstances, and can be expressed through individuals' behaviours, exemplified in the vignettes above. They are listed in the following pages in no particular order, as the importance of each quality will change depending on the circumstances in which the model is applied, with the exception that one quality, 'cohesion', trumps all others in importance.

Cohesion: We Demonstrate Mutual Respect in Practice

'Mutual respect' reflects reciprocal pathways of interpersonal behaviour that recognise and value the input of the other, throughout a system. Cohesion is different from teamwork. Teams that can maintain professionalism and have a clear hierarchy can prosper. But this is not mutual respect; such teamwork does not promote resilience, as the team has nothing to rely upon if the professionalism or order is challenged, as in a crisis.

Respect is promoted by understanding. A challenge for a complex system such as health care is the difficulty one profession or craft group has understanding another. Often the goals of different groups or individuals may appear to be conflicting. Resilience in health care relies on the agents of the system gaining an understanding of the roles of other agents and how only the complete system is able to deliver optimal care. This is the essence of the discussion of the myth of the sharp end/blunt end at the start of this chapter.

In many ways, professional relationships resemble personal relationships: they work best with clear goals; they rely on communication and succeed or fail on the level of trust and respect felt and displayed.

Differences, can, do and should occur, and need to be worked through in a safe way. Breakdown of professional relationships and disintegration of clinical teams can and do lead to sub-optimal patient care. This implies the need for negotiation skills and practice of give-and-take. Understanding the interests and values of other agents is the first step to establish common ground and develop relationships, thereby enabling negotiation to create value within the system, rather than simply to distribute value to the agent(s) with the greatest power. In a complex adaptive system (Clay-Williams and Braithwaite, 2013), it is the influence of negotiation that gives direction to emergent behaviour.

We often say that 'you don't have to like the people you work with, you just have to work together', and we suggest that is a manifestation of work-as-imagined (WAI). To get to the point of mutual respect, you actually need to find something to like about each other, recognising, accommodating and compensating for weaknesses. This can start with small things, and build over time, it may be as simple as 'I like that they turn up on time'. Our juggling nurse manager described earlier draws on a long relationship of mutual respect and trust with her senior doctors to get the co-operation she needs. The WAD of mutual respect requires conscious effort and persistence; it does not just evolve of its own accord.

Capture: We Know What Is Happening and We Know What Is Coming

Capture may be described as "perceiving the world as it really is". This has facets of clinical monitoring, data collection, analysis and reporting, as well as forecasting.

Capture helps create awareness of the situation – that is, an understanding of what is happening around you and what it means for how you need to respond. This is a process to establish an understanding of WAD, rather than WAI.

Capture occurs at all levels of the system. At the senior manager level, capture is achieved through in-depth analysis and understanding of WAD, rather than a superficial reliance on organisational metrics. This requires open dialogue with the people delivering care, discussion and analysis and testing of financial models and operational strategies that then underpin management decisions. The rate of delivery of data needs to match the capacity to process the data into information, avoiding the problem of signal-to-noise disparity where the data drowns out the information requirement and decision-making is compromised as a result. Too much data can lead to information overload.

At the clinician level, such as in our junior doctor example above, this may translate into assessing what the care needs of a patient are likely to be and selecting the appropriate environment for care, as the location of care will determine the intensity of the information available for capture. The clinician then determines the monitoring schedule to match the clinical condition. In a non-critical patient this may be achieved by using fourth hourly observations and a 'track and trigger' tool for early recognition and escalation of care for the deteriorating patient. In more critical patients, it may require intensive 'beat to beat' monitoring in an intensive care unit allowing immediate response. Again, information overload can compromise capture.

In summary, capture is about determining what information is required, identifying how that can be acquired and how it needs to be delivered to inform decisions. It is a key requirement to understanding the balancing point in an ETTO (Hollnagel, 2009a). Capture requires that we recognise the limitations of WAI, and seek to gain a greater understanding of WAD.

Cognition: We Switch on Our Brains and Think, Using the Right Mode of Thought

A great deal of literature has been generated in recent years looking at thought processes. WAI suggests that we need to think through issues, challenges and disparate information to come to appropriate decisions. The reality of WAD is that we simply have too many decisions to make in our everyday life, let alone in a busy day in health care, to allow us the luxury of deep thought; many decisions have to be taken on the fly. Kahneman (2011) has defined "System One" and "System Two" thought. System One is the 'gut' response that is generated when we are first presented a problem. In health care, there are so many decisions to be made on a daily basis that we must rely on System One for many of the decisions we make. However, System Two, which requires you to "stop and think", is the only way to work through complex analytical issues and novel problems.

The challenge is to consider which decisions are safe to make with System One thinking, and which require System Two. This will vary depending upon the experience and expertise of the person involved. A more experienced practitioner will be safe to use System One thought processes in far more situations than a less experienced practitioner who needs to operate more frequently in System Two. The decision about which mode of thought requires an act of 'meta-cognition' which can be, in itself a gut response to 'a bad feeling'. We should also consider the various tricks that our brains play on us and how we can make simple mistakes because of the way our brains work when operating in System One mode.

A strategy to overcome such thinking problems is the use of cognitive forcing tools. 'Rule Out Worst Case Scenario' is essential in the ED as clinicians cannot afford to miss a life-threatening illness. It increases the probability that all relevant diagnoses will be considered (Crosskerry, 2002). Such cognitive forcing tools can have a positive influence on resilience.

In many situations, we will make better decisions when we use a combination of System One and System Two allowing a calibration between the two systems to occur, identifying reassuring agreement or discordance which may signal the need for re-evaluation.

Clear Ownership: We Know What We Are Responsible for and We Accept Accountability for Doing It; We Take Responsibility Where We Are Best Placed to Deliver

Clarity of ownership means that all members of a team, including patient and carers, understand their role in care delivery, take responsibility for delivering that component of care and accept accountability for it.

Recognising that you may be best placed to deliver in a critical situation and stepping forward is a critical element. How many times have we seen the 'heart doctor' perform a consult to do an echocardiogram, reporting on the heart which is going fine, but not acknowledging the criticality of the patient surrounding the heart. 'Super-specialisation' of physicians has promoted ownership of an organ rather than a patient. Reliance on such an approach encourages multiple physicians attending to a patient but potentially a lack of ownership of the whole patient. In our experience, this approach may hinder resilience.

At an organisational level, clarity of ownership is both enhanced (WAI) by and hindered (WAD) by structures. Traditional hierarchical structures, usually described in linear diagrams, drive individual accountability to confer ownership for issues that can be completely resolved within the department or division. Conversely, they are challenged to address the more common and complex issues where the required actions need input from multiple departments and divisions. This is often referred to as a 'silo' mentality. 'Ownership' is a complex issue in the WAD reality of a CAS.

Goldratt develops a deep understanding of this phenomenon in his management parable 'The Goal' (1984) when he identifies that 'local optima', where departmental or individual performance is measured by localised performance metrics, fail to encourage broader ownership of organisational (or in our case patient) objectives. For example, the drive to move patients through an ED in 4 hours may look good in performance metrics, yet may compromise capacity for full assessment of patient needs.

We need to own our part in the broader objective, not just what can be measured in our own department, division or doctor group.

Competence: We Have the Curiosity to Seek Knowledge and Skills and Apply Them with a Professional Attitude

A foundational aspect of any health care system is competent staff. In a modern health care system this area is highly regulated. Staff are employed with baseline qualifications, engage in continued professional development, receive peer feedback through performance reviews and processes such as morbidity and mortality (M&M) meetings and regularly submit their practice to audit.

To be truly resilient, so that a system learns, agents of it must teach. This ensures knowledge transference to others within the system and sustainability for the system. It is an often overlooked but core element for resilience. This transference is in terms of both technical and non-technical skills. This requires a shift in emphasis to development of non-technical skills, such as leadership and communication, essential for resilience.

Competence is sometimes described as a point in time state whereby a person is able to demonstrate that he or she has attained certain standards that can be measured in performance. We suggest that competence has many characteristics and several stages. Higher-order competence depends on moving past procedural competence to develop understanding of the system, thereby attaining a greater capability to apply procedural skills to improve outcomes. This is described by some as a level of 'worldly competence', knowing the system and exerting positive influence to add value.

Constraints: We Are Aware of the Impact the Environment Has on the Action Required of Us; We Manage Constraints in the Environment to Provide Optimal Care

Awareness of the environment and the particular constraints to quality care that the environment brings with it, are critical to working out what our response needs to be to ensure that we still provide quality care, despite the prevailing conditions.

The environment the system works in can be difficult to define. Is it the bricks and mortar of the clinic, or is it the political, social and clinical forces at play at any given time? For the resilient system it is both, as the challenge of optimal system performance is to manage all these constraints to achieve safe care.

It is not sufficient to merely accept that the environment throws up challenges and constraints, we need to take the next step of identifying how that impacts on what we need to do to maintain safety. An example of this can be described in the context of a remote health service. The limitation on the capability and capacity of staff and facilities is easy to identify. At times, this leads to an attitude of "we will do what we can with what we have". Although this is laudable to an extent, and may indeed be seen to be resilient, it is neither.

To be resilient we contend that you need to look past the constraints and identify what this means for you in practice. In the context of our remote health service, we can modify some constraints once we identify them, and we can adjust our practice to mitigate the risks of those we cannot change.

In our example, we can modify the constraints by improving communication links to major facilities, with real-time videoconferencing input from senior clinicians, a virtual presence in the emergency room. We can also adjust our practice, rather than waiting until it is clear that a patient requires emergency retrieval overnight, we may 'make a call' earlier in the deterioration trajectory and ensure that transfers can occur during daylight hours, safer for patients, pilots and retrieval physicians.

Communication: We Practice the Art of Receiving and Giving Information Necessary to Complete the Task; We Listen to Our Patients, Families and Carers

Communication is perhaps the most touted skill, known to be essential in any system for optimal performance. The resilient system places an emphasis on the skill of listening, for it is only with listening that there can be an appropriate response. True listening can be difficult when what is being said is discordant with what is anticipated. It requires a concerted effort.

Listening to the concerns of the patient and their carers is a key element of resilience.

Establishing a non-hierarchical approach to communication can have a positive influence on resilience, as individuals can feel empowered to speak up when they see something is wrong.

A health care system is a truly multilingual organisation. Notwithstanding the different cultures and languages of the people within it; the professions within such a system engage in disparate dialogue with different vocabulary and gestures.

Clinical staff have their jargon while managerial staff have theirs, and often a common acronym or expression will have quite different meanings across different groups.

One of the best examples of this is the use of the word 'barrier'. For a manager, a barrier may be seen as a positive thing, a technical word describing a protective input in risk management. For a clinician, however, a barrier is an obstacle to good things happening.

Patients will struggle to understand dialogue based on a medical vocabulary. The resilient health care system recognises this and encourages forms of communication to work against such obstacles and provide the opportunity to test and confirm understanding.

Culture: We Put Care at the Centre of Everything We Do; We Understand That People's Responses Are Learned from Past Experiences; We Demonstrate Compassion

We need to show that we care through our behaviours.

If we put patient care at the centre of everything we do and demonstrate compassion, and we show this through our behaviours, we will influence the attitudes and the behaviours of others around us, creating a positive culture in a system that perpetually reinforces itself, a resilient system.

Culture is often seen as an amorphous cloud that surrounds us, impossible for the individual to control. While control may be elusive, influence is not.

Once we recognise that we may only have a tight locus of control, but a broader sphere of influence, we can recognise how our individual behaviours impact on others and create a resonance that helps to change the culture around us.

Challenge: We Demonstrate Confidence in Knowledge and the Courage and Conviction to Act to Achieve What Is Required of Us

One of the most common findings in investigation of adverse events is that someone, somewhere in the process of care had seen that 'things weren't right', but had not raised their concerns, either because they were not sure of their ground, were concerned about managing relationships or did not feel that they had the authority to challenge. WAI suggests that calling for help is easy and common. WAD realises that this is far from true. Challenge requires courage and acceptance, a common language and recognition from all parties that it is not only allowable, but also necessary. It is an uncomfortable truth in resilient health care.

This is as much an issue in management of health services as it is in direct delivery of patient care.

In the resilient health care system, individuals are equipped with tools to assist in challenge, and a language to support it. An example of where this is commonplace and accepted is aviation, where a structured form of language, 'graded assertiveness' allows challengers to gradually and safely move through a standard form of words that gradually escalate their concerns. The people they are challenging are similarly trained and therefore can recognise and welcome the challenge before it escalates to potential conflict. A useful model is *PACE*: *Probe* "I haven't seen it done like that before, can you explain it to me"; *Alert* "Can we check that dose please"; *Challenge* "I think that we are heading in the wrong direction here"; *Emergency Action* "I am going to have to call the boss in on this one".

Compliance: We Follow the Rules That Apply to Us

The reality is that we have many rules that may apply in any given situation. These range from directives, to policies, to consensus statements to protocols, to guidelines. It is often unclear which rules apply in given circumstances, and with which exclusions.

Given the large number of rules and the constant state of flux and renewal of rules, there is significant potential for our rules to be inconsistent with each other. This makes it inherently impossible to comply with all rules at all times. WAI suggests that we make things safer by imposing rules, while WAD recognises that commonly, the converse is true.

Part of the task of the system is to better understand what rules should apply in which form to support good clinical care and to ensure that only those rules that are truly required are in place, and are easily accessible and understood.

There are processes in health care that are linear, and require strict adherence to protocol; this is where rules should apply and need to be followed. The mantra in high-reliability organisations is that "always means always"; to this we could add "never means never".

The Model for Resilient Health Care

We are all familiar with depictions of our solar system and most of us are comfortable with the sense that planets orbit around the sun.

For many of us the high school study of the solar system explored the complex phenomenon of gravity and the forces that are exerted on the planets, and in turn that the planets exert on each other as they move in their different orbits around the sun.

What we are describing in this model is the way that the different qualities within the system interact. Each agent within the system brings his or her own balance of the qualities described in the Ten Cs.

The degree to which they come together and compensate for individual weaknesses through cohesion as a team and contribute their different strengths defines the outcome of our care. We can never be perfect; individual and team performance alters from time to time and we need the tools to compensate accordingly.

The qualities displayed by the system are the amalgam of the qualities brought by the individuals within the system. The degree to which teams of individuals combine their individual qualities, rather than have them compete within the teams, is defined by the quality of cohesion – hence cohesion may be considered to enhance the force of the other qualities. That is, cohesion is a 'force multiplier'.

The qualities of the system will exert a 'gravitational' force on each other. This can have a positive compensatory effect, or in some situations can lead to dysfunction and decompensation.

Understanding this helps us understand how to intervene to make a system safer.

Can We Teach Resilience?

At face value, resilience as a concept sometimes appears to be a 'state of being', which implies that it is the product of systemic rather than individual forces. We contend that resilience of a system comes, at least in part, from the individual resilience of the agents within the system, and that the resilience of those agents can be enhanced.

The first step in enhancing individual resilience is to understand what behaviours come together to create it; in this model, that is the Ten Cs. So can we teach them?

For many years we have taught aspects of the Ten Cs, in one guise or another. Bringing the concepts together allows us to think about the interplay between the different elements. We can certainly draw from this and teach most elements by detailing the behaviours that exemplify the qualities. In our experience in teaching this to people across the spectrum of health care, the concepts fit with their lived experience of WAD. Most can see how to 'get more of' most of the qualities, but are often stumped at cohesion … how do you get more of that? How do you teach mutual respect? This has taken us some time to understand, and we think we perhaps have part of the answer.

Negotiation

Negotiation as a skill is mostly taught to business leaders, lawyers and sales staff. It is rarely described in health care and even less as an aspect of clinical practice.

In teaching the art of negotiation, using the techniques of the Harvard School of Negotiation, we first take the students to a better understanding of themselves and how they interact and negotiate with others. They are taught to look for the interests and aspirations of those with whom they are negotiating, integrating those interests with their own to create value rather than taking the default position that "more for you means less for me". They learn how to create "win-win" outcomes. In achieving this, they develop an understanding of individuals, their drivers and their behaviours.

This understanding is a cornerstone of respect, the key to cohesion. Training in these skills helps us to form common intents and goals, fundamental to enhancing relationships. When things start going wrong, the fundamental cornerstone of relationships built on mutual understanding and respect will allow the team to make it right.

Current Applications

The Ten Cs have now been increasingly used in our health service for over 2 years. They originally were developed from analysis of critical events (WAD), and that was the first application in which it was used. Analysis of events using the Ten Cs has allowed us to develop a richer understanding of events that have occurred in a Safety-I framework, and it has also allowed us to examine parallel events where good outcomes have been achieved and adverse events avoided, Safety-II. In turn, this has assisted in much more accurately targeted actions, which recognise complexity of WAD, rather than rely on the linearity of WAI.

Following from this application, they are used extensively in morbidity and mortality review. It has proven much easier for clinicians to explore their practice using a model that recognises the complexity of their ECW and provides them a taxonomy to describe it.

The Ten Cs is now being used as a framework for education about resilience and safety and is increasingly becoming part of the language in our system. The model has highlighted opportunities for targeted intervention to address key clinical issues. For example, changing intensive care handover to a team-based multidisciplinary event has improved communication, capture and cohesion, with apparent measurable effect on departmental efficiency and operational effectiveness reflected in improved access and staff satisfaction.

The conceptual framework of the model has enabled the analysis not only of events, but more importantly of systemic aspects of our organisation, to determine where and what interventions may be appropriate to improve performance. For example, application of the model, and the recognition of the centrality of cohesion, has led to the introduction of intensive negotiation skills training, which has proven revelatory for people from all levels of the organisation as negotiation is a central facet of WAD for everyone and has never been taught as a skill in our system.

Summary

When we look for resilience in everyday clinical work, we find it in abundance through all levels of the health care system.

We need to temper the use of the 'sharp end/blunt end' language as it potentially reinforces a cognitive bias that impairs cohesion across the health care landscape and we propose it is largely a manifestation of WAI.

The Ten Cs model offers a taxonomy of resilient traits seen in the WAD of everyday clinical work. Once recognised they can be reinforced if present or compensated for if absent. In our experience, cohesion is the force multiplier.

If we understand the elements of resilience at an individual level, we can enhance it at the individual level, thereby influencing the resilience of the system.

8

Understanding Resilient Clinical Practices in Emergency Department Ecosystems

Jeffrey Braithwaite, Robyn Clay-Williams,
Garth S. Hunte and Robert L. Wears

CONTENTS

Background

We have begun the task of applying resilient engineering concepts to health care with the advent of two previous scholarly compendiums (Hollnagel et al., 2013; Wears et al., 2015), but there is much work to do to improve our understanding of what happens when things go right in health care environments. This means learning from the under-recognised habituations, caregiving activities and taken-for-granted routines that characterise the ebb and flow of clinical work as it unfolds in everyday practice.

Accomplishing this requires us to chart a course between theories that seek to account for clinical front-line processes and a set of effective empiricisations that can be confirmed or verified by testing. We need to capture and report on salient examples (e.g., Smith et al., 2013) of how, under ordinary circumstances, caregivers take specific actions to ensure that nothing goes wrong.

Introduction

Clinical work is nuanced, multi-faceted and social, and takes place in a myriad of ecosystems. Emergency departments (EDs) are habitats that exhibit promise in helping expose how clinical practices unfurl in real time. In EDs, work is time-critical, richly interactive, idiosyncratically hierarchical and heterarchical and intermittently pressured (Braithwaite et al., 2013a; Forero and Hillman, 2008; Fairbanks et al., 2013). EDs create a trajectory for acute patients, and mostly get things right, despite temporal demands, resource constraints and considerable case mix and workplace complexity (Nugus et al., 2012). Emergency clinicians adjust their work to match prevailing conditions – in quiet periods, at busy times and when crisis occasionally hits – and by interacting with ecosystems beyond their own world.

A simplified version of the complexity of the ED is provided in Figure 8.1. Emergency clinicians provide a flow of treatment in an environment of dynamic interconnectedness (Nugus et al., 2010). They use technologies of support (material anchors and affordances) including computers, pens-and-paper, stethoscopes, medical records, sticky notes, bed allocation boards and referral and discharge letters (Nugus et al., 2011) and the collective intelligence of their combined expertise, relationships and tacit knowledge, applied to regularly presenting, and occasionally unique, case types.

These clinical activities, and the technologies they rely on, accommodate to the various needs of patients who have to navigate through the subsystems of care. The social–professional system serves to regulate, moderate and structure workloads over time, balanced against clinical requirements. Here, at the ED sharp end, resilience has to be continually created. How is this done?

In this chapter, we assess what we have learned from ours and others' empirical work in EDs, and consider a selection of studies that have described

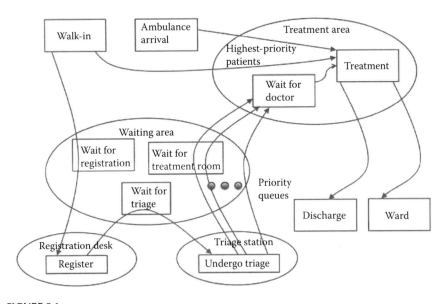

FIGURE 8.1
Simplified systems model of an emergency department. (From Laskowski, M., et al., *PLoS One*, 4(7), e6127, 2009. With permission.)

resilient practices. We draw on five case examples of ED environments, going inside the ecosystems and habitats of the professionals who provide care to patients. Having synthesised the ED behaviours, processes and activities of the cases, we then seek to answer two questions. First, what are the implications of things going right most of the time at the sharp end, which resilience engineering has come to label *work-as-done*, and what is the place for those more distant from caregiving, at the blunt end, who seek to make things safer, but may have a less thorough view of the actualities of patient care, which we call *work-as-imagined*? Second, on the basis of what we now know about ED ecosystems, can we, and if we can, how can we, conjoin the interests of both professional groups?

The Cases

Overview

Nugus and colleagues showed how emergency clinicians meet multiple challenges, and in doing so create a moving 'carousel' (Nugus et al., 2014), balancing their efforts in trading off quality of care for specific individuals with quality of care for more individuals. Sujan and his co-workers (Sujan et al., 2015a) looked at the way handover works in theory and practice, and detected a secondary, unauthorised, secret handover, initiated by front-line ambulance staff and ED ward nurses, designed to keep incoming patients safe. Hunte (2015) analysed an evocative crisis in the form of a riot in downtown Vancouver, Canada, following a sporting event, highlighting the resilience of an ED under rapidly increased demand. Creswick et al. (2009) investigated the networked, micro-structural dimensions of ED clinicians' interactive behaviours, exposing ED tribal characteristics. And Stephens et al. (2015) examined the absorptive capacity and capacity for manoeuvring that EDs can display when backlogged with boarding patients who have not yet found a post-ED placement. These cases create a window into ED environments and their performances.

Case Study 8.1: The Dynamics of the ED Carousel

In trying to understand the way patients flow through the front door of a hospital to the end of their journey (by discharge, admission to another unit or death), empirically minded scholars have all too often taken a statistical-utilisation approach or an input–throughput–output view. Research in this vein has examined topics such as patient demand, access block and overcrowding (Asplin et al., 2003; Coats and Michalis, 2001). Nugus et al. (2014), departing from this type of structuralist, Taylorist thinking, looked at the way EDs function, taking into account social influence, professional boundaries and human agency.

Nugus et al. (2014) constituted the ED as a dynamic system and likened the work activities and patient flows to a fairground carousel (Figure 8.2). This metaphor represents the system as a circular platform,

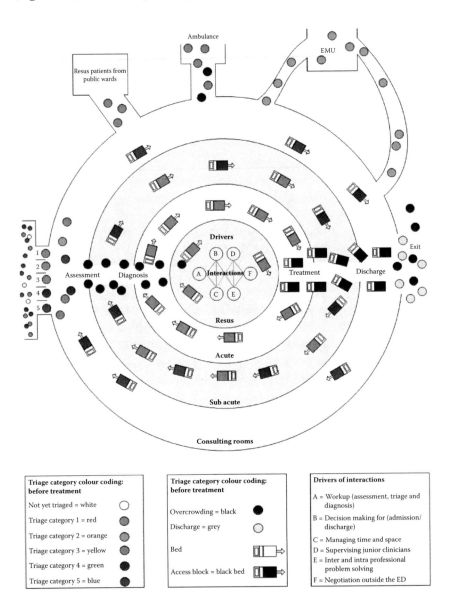

FIGURE 8.2
The carousel model of the emergency department. (From Nugus, P., et al., *International Emergency Nursing*, 22, 3–9, 2014. With permission.)

comparing the individual patients to riders on seats on a merry-go-round, and clinicians to carousel operators. Instead of painted wooden horses mounted on posts moving up and down, simulating galloping, saddles were replaced with beds, and music with clinical and patient decision points.

In the metaphor, the speed of the carousel and tempo of the music, along with the number and type of rider-patients and carousel operators, govern the extent to which the system copes in a process inevitably interspersed with much episodic starting and stopping. The orchestration and choreography of the system are crucial to its success. In order to manage and adjust performance, delivering completed rides and discharged patients, several enabling and constraining factors come into reckoning. Chief among these are the physical layout and design of the carousel, resource allocations, relationships and interactions between players, time made available to each episode, decision-making at key points on the journey and negotiation between stakeholders to reconcile these multiple considerations. In essence, a system – whether a fairground carousel or an ED – needs to spend time adjusting performance and behaviours, matched to changing conditions and circumstances. These are key factors determining how the system achieves its objectives and exhibits resilient characteristics.

A clear point that recurs in this study is the need to reconcile work-as-imagined with work-as-done. In this case, people at the system's blunt end, not actively taking part in daily work – in institutions funding or approving a fairground's operations or inspecting the safety of a carousel, for example, or in health policymaking or hospital management – imagine how work is done. Their knowledge of direct work is once or twice removed from the front line, and the information they receive about it is delayed. People at the sharp end – an actual carousel operator starting and stopping the rides, or the ED clinicians conducting episodic treatment on the clinical front line – directly experience the results of their work (the joy of riders, the pain of patients) but cannot readily see the bigger picture (of other rides and side-shows, or more distal departments or wards). In both fairground carousels and EDs, work-as-imagined and work-as-done are in tension and need constant reconciliation such that the interests of both funders and sponsors of the operations, and those performing work on the line, are met.

Case Study 8.2:   The Secret Second Handover

One major mechanism feeding patients coming into the ED (and getting them onto the carousel ride in the first place) is the ambulance service. Handover of patients from ambulance paramedics to receiving clinicians is a dynamic interplay of interests and responsibilities. Sujan et al. (2015a) noted its multi-dimensional nature in their description of a case study of patient transfers between ambulance services and hospitals in the English National Health Service (NHS). The researchers observed that handover was in the main conceptualised as a formal responsibility between clinical professionals. This transaction included not only the transfer of responsibility for care but the structured communication of information relevant to the patient's condition to the nurse in charge, who would then relay that information to the nurse who would eventually be caring for the patient.

The case study focused on times when ambulances queued in the ED, waiting for patients to be settled and accepted by the nurse responsible for the patient. Although the admission process was explicit, clear and structured, it came with an inherent disadvantage. The ambulance paramedics had to wait when a nurse was busy and could not leave in their ambulance to get back on the road until the nurse completed the handover procedure and accepted the patient. There were various procedural steps that the parties were required to follow in each patient transfer. The onus, then, was to adhere to formalised, work-as-imagined documentation in order to return the ambulance to service quickly.

However, the fieldwork exposed that work-as-done differed from the expected. Paramedics would execute a second handover of patients to the nurse who would be caring for the patient in a partly hidden process standing to one side of bureaucratic procedures. The clinicians on the ground themselves referred to this as the 'secret second handover'. Some more senior nurses knew about this process but did not support it. They thought it was an unneeded duplication. However the clinical front line felt that it improved the flow of communication and the safety of their patients, because they knew the information needs of the nurse in charge and the caring-for nurse differed, and feared important information would be lost in the relaying process.

The secret second handover helped ambulance crews wanting to tell a more complete story and provide more clinically relevant information to the receiving nurses than the normal protocol envisaged. This neatly exemplifies the pressures on people and how they respond under the efficiency-thoroughness trade-off (ETTO) principles articulated by Hollnagel (2009a). The senior nurses were privileging efficiency, and supporting the formal handover, while the paramedics and nurses on the line were favouring thoroughness, and the informal handover, as an additional affordance to good care. That vague clinical unease that comes with being worried about a patient was to some extent appeased by the secret second handover.

This is an example of local front-line practitioners adjusting to circumstances, providing flexibility and increasing the capacity for resilience in the everyday activity of handing patients over from one point in the system to another (Wears et al., 2015). Emergent adjustments to performance, and making dynamic trade-offs, are characteristics of complex adaptive systems (Braithwaite et al., 2013a; Robson, 2015). They serve to meet local, variable needs and resolve tensions between formal aspects of work and informal processes that manifest to get work done effectively.

From this case study, we can see that communication as dialogic sense-making appears to be one of the keys to resolving the differences between work-as-imagined and work-as-done. The secret second handover used a commitment to communication to explore and tackle differences in the mental models. The dissonance and unease, evident when work-as-imagined and work-as-done differed, was attenuated.

Case Study 8.3: Handling a Riotous Infusion of Patients

Although any individual patient's arrival at the ED (whether or not by ambulance) is not predictable, overall expectations of patient numbers at various times of day are usually within foreseeable limits. Sometimes, however, EDs are swamped by unexpected increases in volume.

In 2011 in the normally orderly Canadian city of Vancouver, British Columbia, an important hockey game came to an end with the loss of the Vancouver Canucks to the US Boston Bruins, a rival team. Ice hockey is the national sport of Canada, and this match was high stakes: the Canucks and Bruins were contesting the Stanley Cup, the game's most prestigious trophy. Signifying its importance, this final event of the series was one of the most watched-ever matches in Canadian history.

The population in Vancouver who are sports fans (a large proportion of the city) saw the result of this match – the Canucks lost 4–0 but they only lost the seven game series 4–3 – as an injustice. Multiple catalysts for disorder were present: 150,000 people were watching the game on two large television screens in a two-block city 'fan zone'. Up to 500 people had been arriving every 90 seconds by SkyTrain alone. Alcohol consumption was rife and the mood was boisterous. Before the game ended at 7:45 p.m. mild fighting and bottle throwing had broken out.

A phase transition is a hallmark event in complexity theory: it describes a change in the population from one state to another, in this case from 'fans' to 'mob'. The police began to respond, and shifted from a "meet-and-greet" approach to dressing in riot gear by 8:00 p.m. By then the fan zone was becoming extremely congested and signs were accelerating that a riot could be building. Once the shift from passionate support to outright riot was underway it rippled through the crowd. Feelings of being robbed (despite the decisive scoreline), discontent and alcohol-fuelled violence now radiated through the populace: police cars were set on fire, shops looted and property destroyed. Police began countering with increasing force and tear gas.

The ED at St. Paul's Hospital, the nearest available facility, received in the order of 147 injured people and their supporters, most within 4 hours between 9:00 p.m. and 1:00 a.m. the next day. These were recorded later as the hospital's busiest-ever hours. Of the patients seen that evening, four in five came from the riot. Multiple other patients not recorded were hosed down for tear gas exposure or treated at the first-aid station just outside the ED.

The ED responded with a well-prepared plan for dealing with a crisis of this kind, but this eventuality sorely challenged the capacities and resilience of the staff. Largely, despite extremely high workloads and challenging patients, the ED coped well. Its systems were tested, but its clinical capabilities and propensity to manoeuvre in a short period of time all came through with flying colours.

Hunte (2015) attributed the resilient performance, applying the resilience model of Hollnagel (2012b), to prior learning in response to previous crises, anticipation and preparation, ongoing monitoring as the influx of patients increased, and adjusting behaviours by way of response. In a real sense, the ED's staff had that evening taken the crisis planning off the page and into reality.

St. Paul's ED was able to deal with patients coming for treatment despite the rapid increase in workload, the pressures on staff and the intensifying pace of work. Coping factors included the ED's elastic adaptivity, and the trade-offs made in sacrificing lower-order goals for higher-order goals by reducing treatment times and doing essential things as a priority. In these circumstances, paperwork and computer entry tasks waited, breaks were rescheduled and clinicians worked with more focus, reflexivity and mindfulness than normal. Providers from other areas of the hospital and staff from home, seeing a need, came to help. In short, clinicians in the ED on the evening of June 15, 2011, did what they had to do and narrowed the gap between the hospital administration's work-as-imagined (normal circumstances, with a foreseeable increase in workload and activities, expressed in a disaster plan) with work-as-done (a large jump in cases treated in a truncated period). They managed their own internal phase transition in response to the larger, community-based one. The disaster plan formed a structure to work with, but the demands of the event required substantial practitioner bricolage (Braithwaite et al., 2010) to adapt and meet the demanding exigencies of rapidly arriving patients.

Case Study 8.4:   Relationship Structures
Through an Understanding of Networks

In contrast to the carousel metaphor, which is essentially *sociological* in nature, network analysis provides a *structural* window into ED activities by exposing the patterns of organisation and webs of relationships between individuals. Creswick et al. (2009) investigated 103 staff in an ED, examining participants' reports of how they interacted when they were problem solving, seeking or giving advice about medication safety and socialising with others. These studies illustrated how the network's structures and the relationships between clinicians altered depending on the questions asked.

If people wanted help with a problem they were facing, or were sought after by others to assist with someone else's problem, they would do so almost exclusively with members of their own professional grouping. Similarly, participants would seek advice from their clinical colleagues in their own professional cohort if they had a medication safety problem. Socialising, too, was heavily professionally based.

The overriding finding was that individuals were much more closely connected to colleagues *within* their professional groups than *across* them. This study of course was not the first to show how tribal clinical professionals are when they interrelate *in situ*. However, this work illustrated how the within-professional interactions are deeply embedded in the micro-structural layers and heterarchies of the ED. Figure 8.3 illuminates this in the case of the problem-solving network.

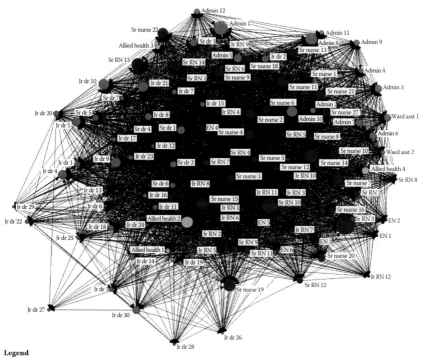

Legend
Jr dr, Sr dr: junior doctor, senior doctor
Jr RN, Sr RN: junior registered nurse, senior registered nurse
Allied health: allied health staff member
Admin: administrative staff member
Ward asst: ward assistant

FIGURE 8.3
Problem-solving network in an ED. (Adapted from Creswick, N., Westbrook, J., and Braithwaite, J., *BMC Health Services Research*, 9, 247, 2009. With permission.)

A lesson from this study regarding the distinction between work-as-imagined and work-as-done lies with the challenge of multi-disciplinary care. Most stakeholders would argue in favour of greater levels of teamwork (Clay-Williams and Braithwaite, 2009; Freeman, Miller and Ross, 2000) and inter-professional practice among their clinical cohorts (Braithwaite et al., 2012, 2013b). According to this logic, greater levels of teamwork and inter-professionalism would be advantageous, more productive and safer for patients. But the reality on the ground in health care, which every clinician experiences, is that hospital services are divided along professional lines. The trick is to preserve the benefits conferred by having concentrations of like professionals (contributing key things such as within-profession comradery, shared interests, support structures, and aiding in the training of future health professionals) with sufficient teamwork and inter-professional interaction to ensure patients do not fall through the system's cracks, and fail to benefit from cross-fertilisation of ideas and synergies.

Case Study 8.5:   Patient Boarding

Although everyday activities are not as dramatic as the events reported in Vancouver in case study 8.3, EDs have to cope all the time with transient patients, variable throughput, temporary surges in patient populations and rapid emptying of patients from their domain of responsibility. The carousel models and the Stanley Cup riot, exemplified above, are specific case instances of how they do this under differing circumstances, and the network analysis of Creswick and colleagues reminds us of how tightly bound within-professional groups are in accomplishing their activities.

But a grey area is at the point where patients complete their immediate treatment, have not been able to be discharged from the ED into the community or home and require admission to the hospital. In the carousel model, this is where patients stop their ride and go on to another part of the fairground. This pivotal juncture in the ED system is where patients are required to 'board' before they can be accepted by an appropriate inpatient unit.

To apprehend this well-known systems problem, accounts have relied on complexity theory, systems dynamics models and concepts such as adaptive capability, capacity for manoeuvring or systems bottlenecks or blockages (Asplin et al., 2003; Coats and Michalis, 2001; Forero and Hillman, 2008). In this fifth case, Stephens et al. (2015) followed patients who needed to be referred to either intensive care or a mental health unit after ED treatment. They showed, following Nugus et al. (2009), how the ED clinicians had to package and 'sell' patients to the relevant inpatient department (see also Hilligoss, 2014; Hilligoss et al., 2015). The receiving units had to be willing to take those patients. The Stephens' research team carefully documented the negotiation skills and delicate workload trade-offs that were at the core of whether an inpatient unit would take a patient from ED, and under what circumstances.

Two kinds of adaptation became apparent in the boarding patients' problem. It seems clear to us that there are common organisational and clinical routines and repertoires exhibited in the performance of EDs. Clinicians have constraints on their organisational freedom because of policy, procedural and managerial requirements. Yet at the same time they have extensive clinical degrees of freedom. When needed they exhibit considerable capacity for manoeuvring such that ED systems can cope with quite pronounced surges in demand (case 8.3). The second accommodation was by the inpatient unit. They are typically full or close to capacity, and would have to make a bed available for an incoming ED patient by discharging another patient under their care. This required delicate decision-making in order to trade-off a discharge for a new patient.

Collaboration was at the heart of the relationship between the ED and the relevant inpatient unit if a successful transfer was to be made. When the contrasting negotiation stances were analysed, it was clear to Stephens et al. (2015) that there were defensive behaviours from both receiving and transferring units, and local levels of adjustment and accommodation. There were sometimes direct and sometimes more subtle aspects to the negotiations, and, depending on circumstances, variably weak and stronger attempts at reaching a collaborative agreement concerning specific transfers. Much of the negotiations were done by telephone and, especially if more important or urgent, occasionally face to face.

Key constructs in explaining resilience from this case included the extent to which mutually constituted agreements could be reached (i.e., how to achieve a win-win with no one losing face). The parties proposing and eventually accepting a patient for transfer were involved in an ebb and flow of interests and the orchestration of work. There were multiple tactics and strategies deployed in proposing and accepting the movement of patients, and adaptive behaviours in support of individual agents and groups to achieve their goals.

Discussion

We have described various embedded facets of resilience in complex ED ecosystems. To do so we drew on case studies articulating aspects of how patients can gain entry into the ED system (case 8.2), proceed onto a merry-go-round of activities (case 8.1) in which they are treated by clinicians organised primarily into tribes (case 8.4), with many patients proceeding through boarding to eventual placement elsewhere (case 8.5). Within the ED, we highlighted how there are common organisational and clinical repertoires, that when called on clinicians have limited organisational but extensive clinical degrees of freedom, and that when needed there is considerable capacity for manoeuvring such that the ED system can cope with quite pronounced surges in demand (case 8.3).

The ED ecosystem is complicated and intricate (see, e.g., Figure 8.1 and case 8.1) but with some core, recurring functions (Figures 8.1 and 8.2), features (e.g., cases 8.2 and 8.5) and structures (e.g., Figure 8.3 and case 8.4). It has devices to mediate demand, create buffering and contribute to making care safer for patients such as the ability to manoeuvre (case 8.3), make informal arrangements for care (case 8.2) and mobilise capacity to conclude new arrangements through bargaining, and making trade-offs.

In such an environment, levels of trust are needed, both within the ED, and with its external partner units and departments, to help make things work. Multiple times, reciprocity – tit-for-tat-type behaviours – were observed (exemplified in cases 8.2 and 8.5, for instance). It is hard to see EDs succeeding without the *you-scratch-my-back-and-I'll-scratch-yours* phenomenon at work across busy shifts, days, weeks and months. If staff within ED do not get on collegially, and in turn fail to forge cordial working relationships with adjacent clinical units, their effectiveness and efficiency are likely to be compromised. Trust and reciprocity, then, are ingredients born of necessity in making for a resilient environment.

To build an understanding of this in more depth, we return to our initial questions. What are the implications for resilient health care, and the place of the blunt end, in the milieu we are describing here? The case features we have drawn out turn on complexity theory. EDs are rich examples of complex adaptive systems (see Braithwaite et al., 2013a and Robson, 2015, for extended analyses). Complex systems exhibit features such as hierarchy and heterarchy, and in places like EDs comprise networked agents (who can be relatively tightly or loosely connected, or somewhere in between) whose actions produce behaviours that are said to be emergent, i.e., which display as patterned regularities. They have cliques, subgroups and nested subcultures.

Complex systems adapt and alter longitudinally, and the constituent subsystems and subgroups change at differential rates and on varying timescales. They are hard to lead, and are rarely amenable to simplistic, linear 'solutions'. Adaptation and accommodation to the prevailing circumstances, and trade-offs, deals and negotiations, are occurring all the time. The occasional phase transition is experienced, where the system temporarily or permanently changes form, such as in the Vancouver riot example. The propensity to manage this habitat is always limited; improvement initiatives need to be flexible. The complex adaptive ED system, as we have seen, is all too often stretched, and displays attributes of resilience or brittleness, depending on the capacities in the system, the kinds of patients encountered and the ebb and flow of workloads.

On the basis of this understanding of complex environments, and the nature of the ED described in the cases, how readily can we square, conjoin or reconcile sharp- and blunt-end interests? Clinical work of the type we have analysed here is fluid, textured and changeable. It can only be accomplished if people treat patients by continually adjusting their care to the prevailing conditions. It is not obvious how people doing clinical work could alter it to correspond to all the policies and procedures mandated or suggested by those at the blunt end – assuming it was possible readily to effect change to entrenched clinical cultures, historical practices and ways of working. Alternatively, it is hard to visualise those at the blunt end altering the landscape of policy and procedures that they have painstakingly developed, funded and sanctioned to match more accurately the clinical care delivered by the front line.

However, our case studies have revealed some clues as to how we might reconcile work-as-imagined and work-as-done. These turn on, and can be detected in, organisational characteristics such as communication, negotiation, teamwork, trust and shared dialogue. Collectively, these are the stuff of organisational resilience, and are at the heart of how people with one set of views (working separately from the front line) and another set of views (working on the front line) can meet in the middle and share their contrasting models of sense-making.

We need to encourage more cross-fertilisation of viewpoints, ideas and understanding, then, across the dissonance between the sharp and blunt ends. Yet the world at the coalface is characterised by uncertainty, complications and unpredictability, and the world once or twice removed is characterised more by linear frameworks of understanding, predicated on input–process–output thinking. Essentially, they have world views that are poles apart. The Germans have a very good word for the totality of one's perspectives: *weltanschauung*, comprising the elements *welt* ('world') and *anschauung* ('view', 'outlook'). The essential *weltanschauung* of each group contains its overarching mental models, cognitive orientations, epistemological standpoints, values and praxeology, or, more directly, its ways of thinking and doing things. And, quite clearly, these differ. Put that way, altering the deeply embedded, distinguishable world views of those doing work and those imagining it, simply put, cannot be easy. The tensions and paradoxes surfacing in the cases cannot be resolved structurally or by *a priori* rule-making, but are better addressed by creating spaces for discussion and negotiation among the competing world views (for a further discussion, see also Wears et al., 2006).

Perhaps we can look elsewhere for inspiration, because as hard as this problem is, it does happen in other areas of life – at least, to some degree – where differences are even more pronounced. Political parties, in rusted-on opposition to each other most of the time, sometimes get to a bipartisan position. Companies such as Zappos, the American online shoe company experimenting with holacracy, or radical democracy between managers and staff (Anonymous, 2014), and those organisations following cooperative models, have been sometimes able to narrow the distance between managerial and front-line interests, despite their profoundly different mental models.

It is not only organisations and political parties that can provide lessons to help bridge the *weltanschauung* of the managers and the managed. Perhaps we can learn from the reconciliation processes of countries like South Africa following apartheid, and Rwanda in the light of the 1994 genocide. South Africa had its Truth and Reconciliation Commission (Wilson, 2001) and Rwanda its National Unity and Reconciliation Commission (Zorbas, 2004), engaging across time with restorative justice and promoting ongoing, fruitful exchange, even after the most difficult historical episodes of enmity and complete disengagement. These examples are much more dramatic than the problem of reconciling those beyond the ED with those on the front line, but they may provide pointers to how bonds are constructed and disparate interests calibrated in the medium and longer term even under horrendously challenging breakdowns.

Unless there can be significantly high levels of trust and reciprocity, a more explicit exchange of perspectives, a willingness of the two worlds to enjoin in improving levels of understanding and a willingness to appreciate the *weltanschauung* of the other stakeholder group, it is hard to see how the divide between work-as-imagined and work-as-done can be connected. Yet if patient care is to be made safer, and caregiving environments more resilient in the ED and elsewhere in health care, finding ways to bridge the two worlds is a crucial precondition.

9

Reporting and Learning: From Extraordinary to Ordinary

Mark A. Sujan, Simone Pozzi and Carlo Valbonesi

CONTENTS

Introduction

The development of a reporting and learning culture is a key feature of successful safety–critical systems (Reason, 1997). A reporting culture ensures that safety management systems (SMSs) are fed with important safety-related information from people who are in direct contact with potential hazards. A learning culture ensures that the organisation is able to draw the right lessons from its SMS, and that the organisation is willing to embrace change when it is needed.

In 2000, the UK Department of Health published the influential report "An Organisation with a Memory" (Department of Health, 2000). The report highlighted the need for the UK National Health Service (NHS) to develop a reporting and learning culture in order to capture systematically data about the extent of patient harm. This recommendation led to the establishment of the National Patient Safety Agency (NPSA), which was tasked with the development of a National Reporting and Learning System (NRLS). Since 2003, NRLS has been collecting data about incidents and adverse events at a national scale in England and Wales. In addition to the NRLS, many NHS organisations are operating local incident reporting systems at the departmental and organisational level. Incident reporting is now well established in the NHS, and it is regarded as a key instrument for improving patient safety and the quality of services (Anderson et al., 2013; Barach and Small, 2000).

While NRLS and other incident reporting systems receive significant amounts of data, questions have been raised about the ability to generate learning, and subsequently change, from this information (Carruthers and Phillip, 2006; Sari et al., 2007; Shojania, 2008). Research suggests that incident reporting data have significant limitations in reflecting the frequency at which incidents occur (Westbrook et al., 2015). Numerous studies have investigated barriers to incident reporting, which include lack of training in the use of incident reporting, usability problems of the electronic systems used for incident reporting, uncertainty about what constitutes a reportable incident, blame culture and fear of consequences, lack of feedback and the absence of learning relevant to local practices (Benn et al., 2009; Braithwaite et al., 2010; Lawton and Parker, 2002; Pasquini et al., 2011; Sujan et al., 2011a).

In this chapter we argue that a key reason why incident reporting fails to lead to sustainable learning in health care is to be found in the concept itself rather than exclusively in any barriers to reporting. With this type of learning the focus is on incidents and adverse events, i.e., on something that occurs infrequently – the extraordinary. The second volume in the present series on *Resilient Health Care* (Wears et al., 2015) suggests that a shift in focus from understanding failures (Safety-I) towards understanding everyday clinical work (Safety-II) is required (Hollnagel, 2014b). We propose that, accordingly, reporting and learning needs to shift attention from the extraordinary failure event towards ordinary, everyday clinical work. Learning from the ordinary might be one way to narrow the gap that exists between work-as-imagined (WAI) and work-as-done (WAD) by providing timely and detailed information about everyday clinical work, and by promoting mindful reflection on practice.

Learning from the Extraordinary

When a patient is harmed, it is a noteworthy and tragic event. Research suggests that around one in 10 patients admitted to hospital will suffer an adverse event, around half of which might be preventable (Vincent et al., 2001). However, it is equally true that most patients will have an uneventful stay, and they will receive care to a high standard. In this respect, adverse events and serious untoward incidents happen infrequently. It is precisely their extraordinary character that attracts attention. Managers, health care professionals and patients need to know what went wrong. Organisations want to ensure that the same event does not happen again.

The focus on extraordinary failures is the key characteristic of traditional approaches to safety management, which leading thinkers in the area of resilience engineering refer to as Safety-I (Hollnagel, 2014b). Safety management from a Safety-I perspective aims to reduce harm and adverse events as far as possible by either eliminating the causes of harm or by controlling the risk associated with these. In order to prevent an undesirable event from repeating itself, the analysis of and the learning from incidents and adverse events frequently lead to the implementation of additional safeguards or defences in order to reduce or eliminate vulnerabilities in the system (Reason, 1997). Such defences often include technological solutions, such as the introduction of electronic prescription systems that have the capability to provide decision support to health care professionals as well as to guard against prescribing errors (Donyai et al., 2008). Further interventions that are prompted by learning from extraordinary failures include attempts at eliminating human error by constraining behaviour and reducing variability through standardisation (Reason, 2000). Examples might include standardised communication protocols or the introduction of mandatory checklists (Hindmarsh and Lees, 2012; Joint Commission Centre for Transforming Healthcare, 2010).

Safeguards, defences and standardisation are examples of well-intentioned interventions that represent instances of formal assumptions about how work should be carried out – work-as-imagined (WAI) (Hollnagel, 2015a). Their primary purpose is to break a particular causal chain in order to prevent a specific failure trajectory from repeating itself. However, the way everyday clinical work is actually unfolding – work-as-done (WAD) – is different, and modern health care systems might best be understood as complex adaptive systems (Braithwaite et al., 2013a). Where WAI is slow to change and relatively stable, WAD is dynamic. WAD is changing constantly because the demands, available resources and environment are changing. Health care settings are full of competing organisational priorities and goals, and suffer from a chronic shortage of resources. Health care professionals are adapting their performance by making dynamic trade-offs between goals in order to translate the tensions and contradictions into safe practices based on the characteristics of the specific situation (Sujan et al., 2015a, 2015b). Such necessary performance adjustments contribute to organisational resilience (Fairbanks et al., 2014; Sujan et al., 2015b). From a WAI perspective, however, performance variability is often regarded as detrimental deviations or violations (Hollnagel, 2015a).

Learning from the extraordinary, from failures and adverse events tends to favour such static interventions, so that events do not repeat themselves. As these interventions are not grounded in a thorough understanding of everyday clinical work, they lack the foundation for appreciating the contribution to the delivery of safe care of the trade-offs and performance adjustments that people make on a daily basis (Cook, 2013a). Instead, such interventions introduce additional constraints, tensions and contradictions, which have to be dynamically managed by front-line staff. The gap that exists between WAI and WAD is reinforced or widened.

Learning from the Ordinary

The thinking behind what is referred to as Safety-II regards the trade-offs and performance adjustments that people undertake on a daily basis as the origin of both success and failure (Hollnagel, 2014b). Health care professionals make context-dependent judgements using their expertise and experience about which goals to prioritise and which specific rules to abide by (Debono and Braithwaite, 2015). Most of the time, these performance adjustments enable successful transformation of WAI into practice; sometimes the performance adjustments are inadequate and lead to failure.

Empirical studies of everyday clinical work, such as those contained in Wears et al. (2015), provide evidence that the focus of patient safety improvements, and health care improvements, more generally, should be on ordinary everyday performance rather than on the extraordinary failure. Similarly, it could be argued that the focus of reporting and learning needs to shift from incidents towards everyday clinical work. Although a deep analysis of the extraordinary might provide interesting and useful insights about the various workplace and upstream factors at play during an incident, it is not possible to adequately describe WAD through the analysis of such incidents alone (Cook, 2013b). Building an understanding of WAD – of the ways in which trade-offs and performance adjustments are made – will only be possible through a study of the ordinary.

For example, such a path to reporting and learning was taken as part of the Proactive Risk Monitoring in Health Care (PRIMO) approach. This approach to organisational learning was developed in order to elicit a rich contextual picture of the local work environment, to move away from negative and threatening notions of errors and mistakes and to encourage active participation and ownership with clear feedback for local work practices (Sujan, 2012). The distinguishing feature of the PRIMO approach is that it focuses on learning from the ordinary, in this case the various hassles that practitioners experience in their everyday clinical work. A brief summary of the PRIMO approach is provided in Table 9.1.

Hassle in this instance can be defined loosely as anything that causes people problems during their daily work. Examples of hassle include, for instance, unavailable equipment such as drip stands on a ward or supporting equipment for undertaking radiographic procedures. There are a number of important benefits of learning from everyday hassle. Among these the most important benefit is arguably that the focus on hassle supports building an understanding of the system dynamics, i.e., of the way performance adjustments are made, and the way work ordinarily unfolds. Reports of hassle typically contain not only descriptions of how the hassle manifested itself, but also how people coped – how they adapted their behaviour in order to continue to provide safe and good quality care (Sujan et al., 2011b). Examples of typical adaptations made by health care professionals include

TABLE 9.1

Characteristics of the PRIMO Approach to Organisational Learning

PRIMO consists of staff narratives, a monitoring survey and long-term and short-term improvements.	Local teams, such as a hospital ward or pharmacy, manage PRIMO for their environment. PRIMO consists of three elements: staff narratives about hassle, a survey to monitor risk perceptions and long-term and short-term improvements.
Staff narratives about hassle document WAD.	Staff contribute free-text narratives about something that caused them hassle during their working week. The local PRIMO champion performs an analysis of each narrative, extracting contributory factors (Safety-I), such as inadequate equipment, and any coping strategies or performance adjustments that staff adopt to manage the hassle (Safety-II).
Monitoring survey is used to build up a risk profile.	A survey based on the contributory factors identified from the staff narratives is distributed at regular intervals to all members of staff (of the local work environment) to build up a risk profile over time.
Long-term and short-term improvements aim to reduce risk, support performance adjustments and prevent participation fatigue.	Based on the survey results, the highest-ranking risk factors are identified as areas for improvement. This is complemented by improvements aimed at supporting the execution of coping strategies and performance adjustments based on the staff narratives. In order to maintain staff participation and to combat participation fatigue, fast and visible improvements ('quick wins') to the local work environment are an important part of the PRIMO strategy that complements its longer-term aim. These are also identified from the staff narratives.

sharing information and personal negotiation to create a shared awareness, prioritising goals and activities and offering and seeking help.

A brief example of a reported hassle from a hospital pharmacy environment is given in Table 9.2. In this situation, the reporting health care professional (the dispensary manager) describes a common situation of arriving at work and finding the dispensary understaffed and under pressure to cope with prescriptions. The dispensary manager goes on to describe how the hassle was dealt with by recognising the seriousness of the problem, by creating a shared awareness of the problem among different parties (such as the drug manufacturing staff and the pharmacy clinical director), by identifying potential additional resources and by revising goals and priorities. In this specific example it might be possible to enhance resilience by making these performance adjustments visible to practitioners and their managers, and by facilitating and supporting such behaviours where possible. A concrete intervention might be the introduction of short morning 'team huddles' aimed at building shared awareness.

When reporting is directed towards the extraordinary, such learning typically is not available. This is because incidents represent situations that have broken down, i.e., situations where the performance adjustments have

TABLE 9.2

Hassle – Staffing Problems in the Dispensary

I came in about 8.30 a.m. hoping to finish off a report from yesterday's (project) meeting, but soon after I arrived a technician told me that the department was extremely short staffed. On top of the planned annual leave, off-site training *and* the three new band 6 pharmacists who needed training up, we had three people ring in sick. I checked the allocation of staff on the rota and went to talk to manufacturing staff to explain that there were no other manufacturing trained staff in dispensary to help them out today. (They were already stretched.) I walked past Clinical Director's office and made him aware that we were really short. Manufacturing leads had not arrived yet so I went to the Distribution team to see if they could have one of the new band 6 pharmacists observe a top up to take pressure off the dispensary so they could focus on getting some work through. I also put myself in for an additional 1.5-hour Accredited Checking Technician (ACT) slot to relieve a senior dispensary technician, so that she could be hands on if needed to maintain some training when the band 6 pharmacist came back.

I realised that my work plan for today would have to be scrapped.

proved insufficient to maintain the system (or care delivery process) under control. However, looking at ordinary, everyday clinical work allows a better understanding of the system dynamics at play and of how organisational resilience is achieved.

Realigning WAI and WAD

WAD by practitioners is necessarily different from WAI in the minds of those who design and manage it. It has been suggested that as the gap between WAI and WAD widens, organisations are prone to becoming more brittle (Dekker, 2006), because policies and procedures that do not reflect the demands and challenges of everyday clinical work might impose additional and potentially contradictory priorities and constraints. As a result, practitioners might become increasingly disconnected from WAI, focusing on practical ways of delivering good quality care and keeping patients safe. Although this creates resilience at the local level, there is a risk that, overall, organisations become more brittle as those responsible for managing work remain unaware of the performance adjustments that are necessary to ensure patient safety (Debono and Braithwaite, 2015). The challenge for safety management and for resilience engineering is to devise ways to keep WAI and WAD aligned. Alignment in this sense does not mean that WAI and WAD are in perfect correspondence. Rather, the intention is to create mutually positive awareness (meaning appreciation) among the different stakeholders of how they perceive work and of how it actually unfolds.

Incident reporting and, more generally, learning from the extraordinary are clearly struggling to achieve this, as their focus is not directed towards understanding WAD. Learning from the ordinary, however, might offer opportunities for creating such a positive awareness and for realigning WAI and WAD.

Learning from the ordinary can provide timely and rich information about WAD in a non-threatening context. Managers usually do not have access to data that directly describe WAD. Their information typically consists of outcome data and other process data that have been aggregated and interpreted over a period of time (Hollnagel, 2011c). Often, these data have been collected in such a way that they can feed comfortably into the WAI perspective, for example data about compliance with standard operating procedures, such as World Health Organization (WHO) surgical checklist compliance data, or data about the frequency of occurrence of harm, such as pressure ulcers. Such data do not provide learning about the positive impact of performance adjustments on successful outcomes, and they might carry negative connotations of errors, deviations and violations. As a result, practitioners might be tempted to hide performance adjustments from those responsible for managing work, and in this way contribute to organisational brittleness rather than resilience (Debono and Braithwaite, 2015). Alternatively, the example of hassle narratives illustrates a potential way of documenting and of learning about how practitioners anticipate and adapt to disturbances through trade-offs and performance adjustments. As hassle scenarios typically represent situations in which disturbances were successfully being dealt with, the performance adjustments that practitioners employ can be brought in a positive way to the attention of those who imagine (i.e., reason about and manage) clinical work.

Learning from the ordinary can contribute to mindful reflection about WAD. Documenting hassle and the way it was dealt with in narratives, and also the subsequent analysis and sharing of this information, encourages an active and ongoing discussion about practice. One of the hallmarks of so-called high-reliability organisations is that they keep the discussion about risk and safety going even in the absence of adverse events (Weick and Sutfcliffe, 2007). With learning from the ordinary such discussions are supported, and their focus shifts from failures and their prevention towards performance adjustments and improvements to practice. In this way, health care professionals can build a better understanding of the consequences of the trade-offs and performance adjustments that they have to make. Frontline staff and their managers can engage in meaningful discussions about continuous improvements in a non-threatening context.

Conclusion

Health care organisations should seek out alternative approaches to complement their established processes for reporting and learning. Current reporting and learning are limited to the analysis of incidents and adverse events. The change that is generated from this type of learning often does not appreciate the positive contribution of performance adjustments, and might widen the gap that exists between WAI and WAD. Learning from the ordinary has the potential to provide valuable information about the judgements and performance adjustments that people make in order to deliver safe care. In this way, learning from the ordinary can contribute to reducing the gap between WAI and WAD by providing rich information about WAD to those who design, manage and evaluate clinical work, and by promoting an environment within which health care professionals can reflect about their everyday clinical work.

Will learning from the ordinary become ordinary, i.e., accepted practice? In the short term this remains doubtful. Following adverse events there is a strong and understandable emotional response among all parties, and such events will always attract attention. However, as the frustration at the lack of progress with reducing adverse outcomes grows, more and more health care organisations might turn to alternative approaches. They might consider complementing their existing safety management practices with proactive approaches from a Safety-II perspective, such as the one described in this chapter.

Learning from the ordinary thus offers opportunities for contributing to organisational resilience, and it is also important to consider that any reporting and learning takes place in a sociocultural context. Further research should aim to understand the factors that enable or inhibit the successful implementation of such an approach to organisational learning. Approaches rooted in realism emphasise the need to understand the mechanisms and the context of change. Further research should aim to identify and to describe the factors that contribute to successful organisational learning across a range of different settings.

Acknowledgements

MS was supported by a research grant from the Health Foundation (Registered Charity Number: 286967). SP and CV were supported by a research grant from the Italian Ministry of Foreign Affairs.

10

Reflections on Resilience: Repertoires and System Features

Richard I. Cook and Mirjam Ekstedt

CONTENTS

We previously described the response of a community care organization to a snowstorm (Eksted and Cook, 2015). The workers exploited existing resources in a novel way to cope with the challenge posed by the storm. The approach they chose entailed giving up some goals in order to increase the likelihood that they would be able to achieve other, more important goals. The ease with which they adapted to the threatening conditions was remarkable. They were able to anticipate how the storm would likely limit their mobility and efficiency. They planned their activities, gathered supplies and began the work quickly. Throughout the day they communicated with each other and their coordinator about the status of their work. They recognized when the work was taking longer than expected and adjusted, in some cases shifting tasks across working groups to allow the more important tasks to be completed by the end of the day.

We make a distinction between the response to the snowstorm – which we called a 'resilience expression' – and the systemic resilience that made it possible. Resilience is an adaptive capacity (Woods, 2015). It is present in the system before the disturbance that evokes the resilience expression begins. Resilience expressions are directly accessible to observers. In contrast, the presence of resilience can only be inferred. Resilience is to resilience expression, as potential energy is to kinetic energy.

In this book, and its precursors (Hollnagel et al., 2013; Wears et al., 2015), researchers have reported their observations of disturbances and responses – that is observations of resilience expressions – as evidence for the presence of resilience. It is not surprising that these observations have been made mostly by practitioners. Much of the resilience in the system requires practitioners for its expression. The capacity to recognize a disturbance or opportunity, to formulate and carry out plans and to adjust work based on the progress (or lack of it!) depends uniquely on the practitioners. They understand how work is done and what impact the disturbance may have on the conduct of work. They also understand *why* work is done and can seek to balance what is desirable with what is possible.

The resilience expressions described in the *Resilient Health Care* series and in other works are diverse. The snowstorm case and other examples have focused our attention on the practitioners' abilities to (1) recognize the specific features of a disturbance, (2) anticipate the ways it might propagate through the system, (3) evaluate and contrast the consequences of these different paths, (4) identify potential responses and (5) select a response that has the greatest likelihood of success and most desired features. These abilities depend heavily on the practitioners' knowledge and experience, and the performances detailed in this series show this clearly.

Several questions arise. How do practitioners develop the repertoire of knowledge and experience needed for resilience expressions? What non-practitioner features of the system are required to support those actions? How are the repertoire and these other features related? Finally, what are the materials and methods that comprise *resilience engineering*?

Practitioner Repertoire

One of the earliest descriptions of what we now call a resilience expression was the Israeli medical system's response to the November 2002 Kiryat-Menahem suicide bus bombing (Cook and Nemeth, 2006). The response was remarkable in many ways. The medical evacuation system quickly removed injured people from the scene and dispersed them to hospitals in the area. The workers at the Hadassah hospital, the largest facility in the region, took on different roles – some identified by donning brightly coloured vests. Usual administrative processing was suspended in favour of swift, definitive treatment by experts. The mobilization and incisive direction of resources was rapid, smooth and effective. Almost as remarkable was the demobilization and resumption of the normal routine; scheduled operative procedures were resumed a few hours after the event.

An observer remarked repeatedly to the surgeons, anaesthesiologists and nurses on how good they were at responding to the bombing. The reply was always the same: 'We wish we weren't!' The implication was that their system had developed its capacity through repeated (terrible) exposure to such events.

The resilience that was made evident by the observed resilience expression had been created over time in response to multiple, sporadic but frequent, distinct but similar events that had been thrust on them. What the observer

saw was a smooth, efficient performance, but this was the result of the evolution of the capacity over dozens of prior events. The resilience capacity was
developed, at least in part, because of the ability to use those experiences
to build the individual and organizational capacity to respond. The smooth
performance was the result of learning. As the old surgical saw goes, 'Good
results come from experience and experience comes from bad results'.

An important element of this smooth performance was its efficiency, in
particular the absence of 'unpurposeful' actions. Resilience expressions
are remarkable at least as much for what does not happen as for what does.
During the trip to the hospital after the bombing the observer's host listened
to the commercial broadcast radio to gather information about the event. He
did not use his cellular telephone to call anyone, explaining that he could
do nothing useful during the trip, that those already at the hospital were
acting in response to the demands there, and that any added telephone communications would be a distraction for them. The emergency room was
crowded with medical and security staff and patients, yet there was little
talking even though some of the patients were in critical condition and others were in obvious distress. Administrators were guiding uninjured family
members out of the clinical area. Operating room staff had already informed
those scheduled for routine operations that there would be a delay while the
injured were attended to. Within 2 hours the emergency room had returned
to its normal activity. Within 4 hours the 'normal' operating room schedule
had resumed.

The development of resilience may thus be analogous to the development of cerebellar control of fine motor skills. Cerebellar dendritic patterns
are initially rich and varied. Practice leads to removal of some dendrite
connections, often described as 'pruning'. At the neuroanatomy level the
effect of learning is as much gradually removing less used (and, presumably, less important) pathways as it is strengthening the more used ones.
The observer's host noted that the early suicide bombing responses ended
with group debriefings but that these had gradually become less useful.
Eventually, the last stages of response became informal exchanges between
individuals rather than 'all-hands' meetings.

We propose that there is a close relationship between the development of
repertoire and the features of disturbances. The resilience expressions take
on a specific character because disturbances have common characteristics
that promote learning across the disturbances. This is a necessary but not
sufficient condition, of course: it is entirely possible to have recurring disturbances but to learn little from them. The effect on resilience may be either
large or small, depending on the breadth of the learning across an organization. In contrast, continuous buffeting by disturbances will prevent learning
either because the variety of disturbances is too great or because there is too
little time to permit the integration of lessons learned. Again analogous to
fine motor skill learning, there are likely to be optimal rates of repetition and
variation.

Domains from which resilience examples have been drawn are mainly those where practitioner training includes prolonged periods of apprenticeship. It is likely that the apprenticeship is crucial both to developing the repertoire for an individual practitioner and for preserving repertoire across practitioner generations. Especially when a particular disturbance occurs so infrequently that an individual practitioner's lifelong experience is likely to include only a few episodes, this sort of transmission may be of particular importance. We note *en passant* that rapid technological and organizational change may undermine the basis for this sort of learning.

System Features

Emphasis on the role of practitioners as active agents in resilience expressions may obscure other sources of resilience. The presence of tools, materials, facilities and information is also a source of resilience. The precise configuration of the workspace, for example, may contribute by placing important tools within sight and reach. Otherwise, insignificant resources (e.g., a flashlight when the main and backup electrical power fail) may take on special significance.*

More broadly, the system may provide or foreclose opportunities for reconfiguration. The recovery of Apollo-XIII depended on both the mission operations team's ability to formulate responses to, among other things, carbon dioxide absorbers that did not fit the working air circulation equipment (Cortwright, 1975). The scheme for assembling the 'jerry-rigged' approach was developed by experts at Mission Control – knowing what materials were in the spacecraft – and built and installed by the astronauts. The contingencies involved in just this aspect of the recovery are remarkable: the presence of experts at Mission Control capable of understanding the problem and devising a solution, the precise inventory of what items were available to be fashioned into the connecting 'mailbox', the ability to maintain radio communications in order to transmit the solution to the spacecraft and even the presence of adhesive-backed tape are systemic features.

Can every component and aspect of a system be considered a source of resilience? Perhaps not, but the resilience expressions compiled to date suggest that almost any system feature might be used in one instance or another. And there are clearly critical design elements that generically contribute to resilience. Although resilience is more than redundancy, the presence of pathways for retreat provides opportunities for goal trade-offs. The

---

* A surgical team working under canvas in Bosnia was able to commandeer battery-powered floodlights from a television news camera crew to complete surgery after an electrical generator failure (Y. Donchin, personal communication).

availability of system reconfiguration to meet changing demands requires practitioner knowledge and experience but also the myriad system 'stuff' that is manipulated.

This line of argument ultimately raises the issue of complexity. The arc of research that leads to resilience begins in the wake of the Three Mile Island nuclear accident with Rasmussen and Lind's report *Coping with Complexity* (1981). A central theme of the report is that the complexity of modern systems, including nuclear power plants, prevents prescriptive definition of operator actions. Since then, complexity has been sketched mostly as nemesis. It is complexity that frustrates operators, blocks diagnosis, creates hidden pathways to failure and obstructs efforts to engineer robust systems.

It is, however, possible that complexity can be an asset to resilience or even essential to resilience. Complexity allows some systems to change goals and behave differently under different conditions. Rather than being solely a liability, systemic complexity provides that requisite variety needed to achieve success in the complex and changing world. From the practitioner's perspective, complexity provides added degrees of freedom that offer opportunities for choice among varied goals and the means for reconfiguration to achieve them. In resilience terms, complexity is, at least sometimes, a resource rather than just an obstacle.

Synergy between Practitioner Repertoire and System Features

A recurring feature of resilience expressions is the close relationship between the specific content of practitioner repertoire and the system features that have major and minor roles in the expression. The availability of specific tools promotes their use and also development of the skill needed to use them. There is a deep ecological relationship between tools and purposeful human abilities. Mumford (1934) distinguishes between machines and tools by the way in which the latter take on the unique characteristics of their users. The refinement of the tool occurs in parallel with the refinement in the user's ability to wield it, both are transformed.

The presence of multiple goals and availability of resources are similarly features that practitioners may shape and *be shaped by*. The practitioner repertoire exploits and is also 'tuned' by the presence of these goals and resources. In the same way that the physical tool is shaped by the practitioner's repeated use, the experience of constructing different response pathways from the available degrees of freedom refines to the practitioner's capability for exploiting these efficiently and precisely. It is likely that system features and practitioner repertoire are linked (perhaps even locked?) together in a mutually sustaining relationship.

Practitioner ability in managing disturbances develops through use of system features. These features are recognized as valuable because of their participation in the expression. The recognition prompts *both* increased appreciation by practitioners of the specific features as sources of resilience *and* incorporation of their uses into the repertoire. Making the features more prominent makes them more easily defended and maintained. Conversely, degrees of freedom that have not been recognized as contributors are candidates for discard.[*]

People who look for resilience 'markers' may therefore find it difficult to identify discrete contributors to resilience. The human performances and system affordances will evolve synergistically so that the individual contributions of one or the other will be entangled. Frequent disturbances will gradually enhance the resilience if the disturbances permit explorations of the available degrees of freedom and support learning. Once again we note that the ability to incorporate these experiences may depend critically on experienced practitioners acting as teachers, mentors and commenters.

Less obvious but certainly no less important is the availability of opportunities to learn about the evolving system. The dramatic events like bus bombings are clearly learning opportunities. But less dramatic situations must also contribute. The 'soft' emergency case (Cook and Nemeth, 2006) is such an opportunity. Learning opportunities are not, of course, all that is required for learning. There may be optimal conditions for learning including the frequency of disturbances, the degree of engagement and the time available for reflection (Cook, 1999). Resilient systems are likely to develop where these conditions exist and are sustained for long periods. Alternatively, where resilience is important, we can expect that these conditions will be promoted and defended.

Possibility of Engineering Resilience

Resilience is currently a 'hot' topic in several work domains including high-tempo medical settings (e.g., neonatal intensive care, labour and delivery, emergency room), air traffic control and network operations. The interest in resilience comes from realization that disturbances are routine and usually well handled in these areas. This is, of course, not news to sharp-end practitioners: in their work, expertise is synonymous with the ability to handle a wide range of disturbances. Ironically the needed expertise is so common there that disturbance handling is treated as 'ordinary' work.

---

[*] To maintain system features not recognized as contributors is difficult. In the MAR knock-out case (Cook and O'Connor, 2005, #87121), the computer system failed and the computer printers that normally printed labels were useless. Typewriters, formerly available in the pharmacy, had been removed and drug labels had to be written by hand. Similarly, failure of a computer-based railroad schedule tool kept agents from providing schedule information because the printed schedule tables were no longer produced (Cook, 1997).

The extraordinary qualities of these performances surface when outsiders look at them through the lens of resilience. From the researcher's perspective, resilience is at least as much a way of looking as it is a characteristic of the workers and the workplace. Ironically, more fully developed resilience is likely to make the work-as-done seem unremarkable and promote even greater separation from the work-as-imagined.

Immediately upon realizing that resilience is important and embedded in the work setting, we are called on to deploy tools to enhance resilience. This is to be expected. Sharp-end people want to increase the degrees of freedom available for handling disturbances and to create more opportunities to control workflow. Those who manage the workplace are disturbed by the vulnerability that resort to resilience implies and want to manage the system resilience directly. Those who finance and oversee these systems recognize the potential economic gain that resilience seems to offer. All these people are pragmatic and conditionally interested in resilience. If it can be enhanced, refined or exploited, resilience has value.

The promise of *resilience engineering* is that systemic resilience can be improved in meaningful, measurable ways by the *intentional* application of specific tools and knowledge. But there is no persuasive evidence that *resilience engineering* is possible. Persuasive evidence would be multiple examples of intentional application of resilience tools and knowledge that resulted in improvements in resilience. The state of resilience research does not allow us to gauge the amount of systemic resilience in a particular system, let alone provide reliable, accurate measures of resilience over time and in response to interventions.

The relationship between measurement and engineering is deep and durable. This poses a risk for resilience. It is quite possible for misguided enthusiasts to promote particular metrics as resilience measures. There are precedents for this, notably *situation awareness*. If the lack of measures is seen as impeding *resilience engineering*, we confidently predict that measures will be developed, 'tested' and promoted. This will be the consequence of the temperature of this 'hot' topic but the effect will be rapid cooling – as the measures are found to be poor gauges of resilience, interest in the topic itself will decline.

It occurs to us that resilience is found in systems in spite of efforts to engineer them. The resilience expressions that we and others have provided depend on some sources that are far removed in time from the expression itself. Resilience expressions are often not formally approved by the organizations. The soft emergency case (Cook and Nemeth, 2006) is successful because of the 'space' around the formal definition of 'emergency' that becomes a region for negotiation. The formal organization relies on crisp boundaries for the term to manage conflicts and assign responsibility and authority. Resilience makes use of a difference that was never designed or purposefully created. The resilience expression is successful because the system is more complex than the rules that seek to manage it.

Resilience may not be amenable to engineering in a conventional sense, and this may be a good thing. Deliberate management of system operations does not give us much confidence that the capacity to engineer resilience would lead automatically to more of it. The ability to enhance something is usually accompanied by the ability to erode it. The desire for greater efficiency and more productivity might well lead to deliberate efforts to minimize rather than maximize resilience, especially when investments in resilience take longer to mature than it takes to realize the gains available from using those resources elsewhere. If resilience engineering is possible, we need to ask who is to be the engineer in charge?

11

Power and Resilience in Practice: Fitting a 'Square Peg in a Round Hole' in Everyday Clinical Work

Garth S. Hunte and Robert L. Wears

CONTENTS

Introduction

Fitting a 'square peg in a round hole' or 'squaring the circle' idiomatically expresses the concept of a misfit or impossibility. Mathematicians since Archimedes have attempted to fit squares into circles in an attempt to calculate π, only to find that the exact number cannot be calculated – π is transcendental and irrational. Nonetheless, it is mathematically possible to fit a square peg in a round hole, provided the diagonal of the peg is less than the diameter of the hole, and also possible, as powerfully illustrated in the story of Apollo 13, where square lithium hydroxide canisters from the command module would not fit the round openings in the lunar module environmental system, to fit a 'square peg in a round hole' with flexible and interdependent thinking ... and duct tape.

It is in this sense of the phrase, that we explore doing the 'impossible' in bridging the divide between work-as-done (WAD) – the square peg – and work-as-imagined (WAI) – the round hole – from the perspective of *power*. Our emphasis is not the power of force and might, but rather the principle that safety is political, and has instrumental value in a bureaucratic organisation (Hunte, 2010). Drawing on Clegg (1989), Foucault (1975), Giddens (1984) and Haugaard (1997, 2002), we take the middle (pragmatic) road between consensual and conflictual views of power, and focus on capacity for action to accomplish everyday clinical work in the context of shared and contested practice.

An Example from Health Care

We ground our discussion in a case where a difference in collective response between emergency physicians and nurses to patients being cared for in an emergency department (ED) waiting room due to hospital access block led to a transformative organisational change. Access block is a safety risk with a known association between mean ED length of stay (as a proxy measure for crowding) and 7-day mortality and hospital admission (Guttmann et al., 2011).

A decade ago, a cluster of patient deaths in the waiting room of an urban, inner-city hospital ED proved to be a tipping point in resilient operational performance. The emotional salience of these stories was profoundly disturbing regarding how the organisation perceived itself. For years, the ED leadership had been pressing hospital leadership about overcrowded conditions, but their concerns had not led to any action to relieve access block. If the hospital was full, then the ED was forced to accommodate both admitted patients (stable) and incoming patients (unknown, potentially unstable). With limited and finite physical space, increasing work demand collected in the waiting room.

From the perspective of one emergency physician (Hunte, 2010, pp. 204–205):

> We're used to running flat out, but then we get three chest pains in a row or somebody who's really sick, then for a brief period of time it's brilliant. People get moved, stuff happens, people are creative … When the chips do get down they pull through and it's almost a joy to be around in that setting because you feel like we're doing some good. Everybody's on the same page and we're working well as a team …
>
> But that doesn't happen on a chronic basis … A bomb has to go off before you can get that sort of cooperation going, and the rest of the time people want to try and make our *square peg fit in the round hole* that's being provided to us … On a given day when the place is in shambles and there's [sic] people vomiting or whatever in the waiting room, to say "well, we're not giving meds" as a blanket statement is really poor. The union can say what they want, and I agree there are certain safety concerns with certain medications, but refusing to start an IV and give someone an antiemetic is not valid at all. That's just being mean. I don't care what rules are around. The emergency department is a different animal from any other hospital unit because we are the interface; we have no control over what walks in the door.

For nurses, there was a perceived risk of harm from an act of commission – administering a medication without adequate monitoring, while physicians perceived a risk of harm from an act of omission – not attending to a potentially unstable patient in a timely way. Both groups of practitioners were attempting to mitigate risk, with their perceptions of safety influenced by metrics of their practice frame works: nursing practice standards and time-to-physician, respectively.

This difference in collective response was a divisive and deciding watershed on the spirit of teamwork and collaboration. Nurses felt uncomfortable administering medication in a space that was not monitored, and there was reluctance to exceed the nurse-to-patient ratio that had been won through negotiation. That effectively left care of patients lingering in the waiting room to the emergency physicians alone. Here, out in plain view, was the polysemous, political and contestable definition of 'safety', as the resilience of operational performance varied depending on the politics of the waiting room.

Yet, out of this tension came innovations in physical structure and care processes such as an Over Capacity Protocol[*] (Innes et al., 2007), Rapid Assessment Zone[†] (Bullard et al., 2012) and Diagnostic Treatment Unit,[‡] as the providers in the department and the organisation moved from a fractious 'either-or' to a more collaborative 'both-and' frame. This process of negotiated power and resistance took time (and remains ongoing), but these novel structures have allowed for significant improvements in the capability to respond, including more timely assessments and interventions, a greater capacity for manoeuvre and improvement in patient flow through the ED.

This is an example of how ED care providers and staff self-organised to deliver care, making trade-offs between competing priorities as they adapted to accommodate clinical demand and balance concurrent care to multiple patients. It also provides a lesson in how resilience in anticipating and recovering from threats to operational performance and safety depends upon the improvisation (*bricolage*) and sense-making of practitioners in dialogic action and relations of power.

Power

Power, from the Latin word *posse* and Middle English Anglo-Norman French word *poeir*, implies the ability to do or act, or to 'be able'. It connotes the capacity and capability to accomplish or influence, and is related to the concept of agency. The term is also related to the Greek *kratos*, as in bureaucracy, and *kybernetes*, as in governance (steermanship).

---

[*] Organisational response to emergency department congestion.
[†] Virtual bed unit for lower-acuity patients.
[‡] Emergency department observation unit staffed by a nurse practitioner.

There is, however, a lack of unity in perspectives on power as a concept, and no single definition of power covers all usage. Power is a 'family resemblance' or language game (Wittgenstein, 1953/2009) concept in political and social theories, with analytical, modern and postmodern paradigms contrasting power as coercion, conflict and control (domination) on one hand, and strategy and consensus (empowerment) on the other. Traditional models of power tend to understand it as something individuals hold (e.g., Weber, 1922/1978), whereas postmodern models emphasise the relational nature of power, shifting the focus to structural relations between individuals, actors and organisations rather than individual actor attributes (e.g., Foucault, 1975).

Modern definitions of agency do not account for situated action within social and ecological relationships. Complexity theory explicitly locates humans within socio-ecological limits (embedded), but omits that these are relations of power. In order to develop resilience, flexibility and adaptability in response to the increasing complexity, interconnectivity, unpredictability and instability of contemporary social–ecological systems, it is necessary to recognise that humans are limited by both ecological relations and relations of power. Agency, thus, is constituted within relations of power, and is an aspect of power relations.

The concept of *practice*, for example, a family of theories of recursive production and re-production of social practice, expresses the mutual interdependence of structure and agency in both constraining and generative senses. Within this recursive duality, human actors perform intentional actions and have the power and 'the capacity to make a difference to a pre-existing state of affairs or course of events' (Giddens, 1984, p. 14).

People produce social systems employing rules and resources during interaction, knowingly or unknowingly reproducing these structures in practice by routines that are generally taken for granted (Hardcastle et al., 2005). Agents are empowered and constrained by structures, both by the knowledge that enables them to mobilise resources, and by the access to resources that enables them to act. Practices, not roles, constitute the mediating moment of reproduction and change in the recursive articulation between actors and structure.

This dynamic relationship is beautifully illustrated in the natural world by the concept of *stigmergy*, or artefact-mediated collaboration (Grassé, 1959). Fundamentally, stigmergy illustrates the recursive nature of structure and action. The concept has been progressively expanded beyond its relatively limited origins in insect societies, and applied to higher-order coordination in economies and human societies (Doyle and Marsh, 2013; Parunak, 2006; Susi and Ziemke, 2001). People take actions based on existing structure; by those actions, structure is reinforced and/or modified, and by those reinforcements/modifications, future actions are influenced.

Changes in formal and informal structures have (un)intended consequences for work routines, the capacity to act and the meaning of work. Work systems cannot match their environments completely; there are always gaps in fitness and a need to adapt. Any system that remains viable over time must be able to cope with unexpected change. It must be able to revise and replace policies and procedures, for variability contributes not only to progress but also to stability in a changing environment. Although bureaucratic structures are often coercive and inflexible, they also enable work performance when they provide guidance and clarify responsibilities without squashing innovation and creativity. The set of constraints and resources channels the set of possible innovative and evolutionary paths. Therefore, resilient organisations appreciate local practice variability as a potential trove of unique innovations and commit resources to their development in order to support and enable adaptive action.

Whereas variable practice is instrumental in maintaining stability amid perturbations, stable mechanisms and limits enable adaptability by providing the background and memory for identifying the unexpected, and freeing discretionary energy and attentional focus. Effective bureaucracy facilitates the transfer of scarce attention and resources from routine to non-routine tasks by fostering trust, reducing uncertainty and providing a framework for emergent action (Farjoun, 2010). This plays out in everyday practice by providing a structure to mitigate against working at cross-purposes when the ED is at overcapacity. By having a graded response to increasing workload, the organisation has the means to distribute the work.

A systemic and collective approach facilitates adaptation by promoting coordination, channelling work in productive directions and guiding and promoting innovation. The duality perspective of practice theories therefore offers insight into how exploitation and exploration intertwine in the messy world of practice (March, 1991; Powell et al., 1996). It reflects a tension that can never be resolved, but must be actively managed (Greenhalgh et al., 2009).

Power in Organisations

Power is an endemic part of organisational life, and usually signifies coercion, the maintenance of hierarchies and subordinated positions, as in the 'rules of the game' (like chess). Political dynamics not only shape rules, but also manipulate practices so that they appear to fit within predefined parameters (Clegg, 1975; Perrow et al., 1972). Politics rearranges relations between people and the distribution of resources through the mobilisation of power (Arendt, 1998). In turn, power is the capacity to influence, a resource to get things done through other people, to achieve certain goals that may be shared or contested. Organisational life and politics are often nasty and backstabbing, but power need not be something necessarily to be avoided. Power can be a positive force, and can achieve great things (Fleming and

Spicer, 2014), like the departmental and organisational changes that have enhanced the capacity to manoeuvre despite increasing workload.

Many contributors to the functionalist and critical bodies of literature adopt a normative approach to power and legitimacy, where power is something that is held and exercised by people and import is placed upon who has more or less of it; they aspire to some idealist view of how power 'should be' in a social system and who should have how much of it. For the rational theorists the ideal is rational, for the critical theorists the ideal is democracy.

In contrast, contributors to the alternative intellectual tradition do not adopt any such idealist starting points. They are not concerned with telling people how power 'ought' to be used; rather they are concerned with studying 'how' power actually is in a social system. For Machiavelli, the achievement of order in organisations does not emanate from a commitment to an ideal form of governance through a benevolent sovereign, but is secured by a strategically minded Prince. Machiavelli conceived power in terms of strategy and practice and as distinctly empirical (Machiavelli, 1975).

Power, then, is a matter of efficacy, the capacity of individuals and groups to achieve their own ends and/or frustrate those of others. In turn, power depends on the availability of resources, which come in many forms, and vary from context to context. In principle, resources offer different ways and means to achieve ends. They may be maximised, invested in and shared depending on context and project. Individuals and different groups of actors will have access to differing resources in differing degrees and in differing combinations.

'Power to' is the master concept indicating capacity for action, while 'power over' is a subset whereby actors are made to do things that they would not otherwise. When actors join together to act in concert (power to), they do so to realise joint tasks, and in so doing they do things that they would not otherwise, compromising and trading off lower goals for the purposes of achieving higher-order negotiated goals like creating safety.

Power and Resilience

Resilience offers a systems framework for understanding how to navigate systems, but ignoring issues of power and inequality provides only a partial view of system dynamics. In situations of uncertainty and change, critical factors for resilience are the capacity to cope and adapt, and the conservation of sources of innovation and renewal. However, interventions with the aim of altering resilience confront issues of governance (steermanship). Who decides? For what purpose and for whom is resilience to be engineered?

Resilience has become a policy 'buzzword' over the past decade despite lack of clarity amid a diversity of approaches and understandings of the relation between the human subject and his or her environment. Whereas 'classical' or neo-liberal views were subject centred, describing inner resources of autonomous individuals for coping and survival against external pressures, more recent 'post-classical' or 'post-liberal' views move beyond the spatial and temporal subject/object divide to understand resilience as 'an interactive process of relational adaptation' (Chandler, 2014, p. 7).

Seeing resilience as dynamic and relational is fundamental to understanding power in an age of complexity. Focus on self-organising adaptation offers an alternate against hierarchy and anarchy by enabling governance through the reality of complex life rather than 'over' or 'against' it. Here, we understand governance or steermanship as the manner in which power is exercised. As an alternate to assuming a system characterised by stability and equilibrium, emphasis is placed on understanding change and surprise (non-linearity and emergence) and how governance arrangements cope with and adapt to cross-system interactions (Ostrom, 2007) in a dynamic and changing environment (Duit et al., 2010).

Governance

A paradigm based on planning for efficiency, standardising for easier social control and reducing variability has come to pervade bureaucratic practices. Problems are framed as technical and administrative challenges devoid of politics. With good information and technical skills, the future can be blueprinted. This is work-as-imagined in the orthodox ideology of authoritarian high modernism (Wears and Hunte, 2014). However, this view of the world has been challenged again and again by practical experience (Ostrom, 1990, 1999a, 1999b) that was gained from work-as-done. Uncertainties and non-linearities often arise from both complex internal feedbacks and from interactions with structures and processes operating at other scales (Gunderson and Holling, 2001).

Complex adaptive systems demonstrate capacity for self-organisation, adaptation and learning. Self-organising systems are able to buffer impacts and do not need to be continually invested in to persist (Holling, 2001; Ostrom, 1999a). The ability to learn and adapt implies that a system can get better at pursuing a particular set of management objectives over time and at tackling new objectives when the context changes (Folke et al., 2005). The capacity to cope with non-linearities or other forms of surprise and uncertainty requires openness to learning, an acceptance of the inevitability of change, and the ability to treat interventions as experiments or adaptive management.

Specific governance propositions to support resilience that have been explored include (1) participation that builds trust, and deliberation that leads to the shared understanding needed to mobilise and self-organise; (2) polycentric and multi-layered institutions improve the fit between knowledge, action and context in ways that allow societies to respond more adaptively at appropriate levels; and (3) accountable authorities that pursue just distributions of benefits and involuntary risks enhance the adaptive capacity of vulnerable groups and society as a whole.

Polycentric and multi-layered institutions appear to be important to building or enabling the capacity to manage resilience. Local governance arrangements can develop to better match the varied social and ecological contexts and dynamics of different locations. Local monitoring may provide effective early warning systems, and monitoring of interventions allows safe-to-fail experimentation. Local knowledge can inform local actions in ways that a single centralised system cannot, and yet decentralisation without corresponding accountability may reduce the capacity to manage resilience. Upward and downward accountability is a safeguard that prevents capture of resources and provides ways to reorganise after failures.

Discussion

The case example offers a story of success, but there are myriad examples of naked assertions of 'power over' where resilience is lost amid bickering and posturing. The ability to respond to patients in the waiting room was constrained by organisational and professional rules. Yet, as the problem of ED crowding demonstrates, the complexity of clinical work goes beyond what individuals know and exceeds what can be formalised as knowledge and rules. Everyday practice harbours unpredictabilities that require different solutions than those provided by the bureaucratic paradigm that privileges formal knowledge. This perspective recommends a dialogic approach that moves beyond the traditional dualism of 'top-down' and 'bottom-up' into a generative partnership between leadership and practitioners that overcomes the limitations of strategies that fail to account for practice and the prevailing conditions within health care organisations, and initiatives that fail to connect with system-wide learning.

It is difficult to avoid power when attempting to apply the concept of resilience to questions of governance. In exploring the resilience of systems, we must ask not only the resilience of what, to what? We must also ask for whom? (Flyvbjerg, 2001; Lebel et al., 2006). This is essentially a normative issue requiring judgement or *phronesis* (practical wisdom). This perspective invites us to consider issues of change and stability, adaptation and design, hierarchy and self-organisation in the study of multilevel governance systems. Moreover, in addition to traditional benchmarks such as efficacy, accountability and equity used when assessing public governance, a resilience perspective on governance would also consider sociotechnical interactions, maladaptations, vulnerability and innovation capacity as integral parts of evaluating a given governance system (Nelson et al., 2007).

Diversity and decentralisation increase the capacity of governance systems to handle complexity (Ashby, 1956). Diversity is necessary in order to develop the flexibility and creativity that are needed for systems to adapt to unpredictability and instability. However, which type of governance system has the largest set of available actions remains an empirical question. A governance system consisting of diverse and flexible semi-independent networks may have a larger set of action alternatives, but it may also have a more constricted repertoire of action, through lack of coordination and fragmented communication. Complexity management further requires governance systems to be both flexible and stable at the same time (Voß et al., 2006). Flexibility is needed to adapt to novel and unexpected circumstances, but stability is of equal importance for ensuring that governance systems provide a predictable arena for interaction between actors.

Perspectives on power and conflict are necessary in order to give an account of the dynamics of everyday work (Antonsen, 2009). Introducing a more power-oriented view on safety and resilience within a complex adaptive system also serves as a basis for ethical considerations (Cilliers, 1998) regarding the improvement of safety. Furthermore, moving from a fractious either–or to a more collaborative both–and frame invites consideration of the other (Bakhtin, 2010) in dialogic relationships of negotiated power, offering a generative interdependent partnership between leadership and practitioners in bridging the WAD–WAI gap – and fitting a 'square peg in a round hole'.

Part III

Methods and Solutions

Jeffrey Braithwaite, Robert L. Wears and Erik Hollnagel

Which methods work best in uncovering resilient performance? How can we best appreciate expressions of resilience? What can we learn about it from research methods that develop an understanding of work-as-imagined (WAI) and work-as-done (WAD)? When all is said and done can we, and if so by what methods can we, reconcile WAI and WAD? To what extent can the conduct of research or well-designed evaluations help provide solutions to some of the problems we have encountered to date, and make good on the applications we have outlined thus far? It is to such questions that we now turn, in this final offering of four chapters under the heading Methods and Solutions.

In the previous two *Resilient Health Care* volumes we learned much about methods and solutions people have used to uncover, work *on* and work *with* resilience in health care settings. At the sharp end, as we are keenly interested in seeing and understanding the resilience that is expressed in the behavioural flows of groups and systems, we should pay attention to both the formal and informal practices that make up normal, everyday clinical work. Of course, much the same might be said for apprehending what goes on at the blunt end where people attempt through their activities to prepare, organise and manage everyday clinical work. Chapters in all three *Resilient Health Care* books have ethnographically examined WAD – in wards, intensive care units, emergency departments and community settings, for example, and have provided a solid platform for appreciating WAD. To complement these, more ethnographic descriptions are needed of, for instance, how blunt-end plans, tools and methods are selected, prioritised and approved, and then get funded, made, modified, approved for use, applied or subsumed into clinical processes, and then reviewed and evaluated, or in other words, how work is done at the blunt end. This would be of critical value in completing our picture of how (or the extent to which) resilience manifests.

Creating ethnographic profiles both of how work is done and of how work is imagined by people at the sharp end and the blunt end alike is necessary, but not sufficient for a more thoroughgoing appreciation of how health care is resilient. To appreciate coalface or managerial intricacies, a range of social science methods can be applied as a complement to the ethnographic descriptions that have been presented in this and the previous *Resilient Health Care* volumes. These methods include case study accounts, interviews, attitude surveys and focus groups. Specific tools can be used or modified in conjunction with these data-gathering approaches – using social network analyses, for example, or the functional resonance analysis method (FRAM).

Statistical modelling of quantitative data can also be undertaken. This can be done as part of specific studies, but health systems create huge data sets – not just for accidents and incidents, but admissions, discharges, patient outcomes, deaths, drug administrations and the like. This has come to be called 'big data'. These stores of quantitative information can give additional insights into systems' performance that complement qualitative accounts, and multiple data sources can be linked to appreciate how work is managed, how care emerges and how resilient practices are influenced and expressed.

Regardless of which research approach is chosen, all methods and solutions come up against a core problem. Resilience itself cannot be seen, any more than power or complexity can. Resilience is not something a system has, but something that it does (or the way that it does it). Resilience, power and complexity are not objects we can inspect unproblematically in the way we can examine chairs, or computers, or X-ray machines, but are instantiations in the performance of a system. Put another way, following Cook and Ekstedt's point in Chapter 10, resilience is not directly accessible, but its expressions are. So whatever method is selected to study resilience, it will need to add to descriptions or understanding of what happens behaviourally when people accomplish clinical activities, or set standards, rules, policies and procedures for them by which to influence those clinicians' accomplishments.

Over and above classic social science methods, an additional way to appreciate resilience is to model the aspects of the phenomenon we are trying to describe and understand. Models are always simplifications of reality, of course, and as abstractions of the world, they always fall short of complete descriptions or understandings. But they can illuminate important variables worthy of attention. In Chapter 12, Anderson, Ross and Jaye mobilise a model of their own creation, which suggests WAI represents two key concerns – demand and capacity. WAD, balanced between demand and capacity, feeds forward into two other key outcomes – success and failure. This model points to a challenging question: how do clinicians get it right when there are always palpable risks of failure, when they are under vigorous demand and capacity pressures and when there is constant misalignment of demand and capacity? For Anderson and colleagues, this problem, framed in this way, can be studied qualitatively via ethnography, interviews, focus groups and the like, and quantitatively, by measuring demand and capacity, and taking metrics of success and failure, via direct and proxy measures.

A second example, and potential solution for reconciling WAI and WAD, lies in simulation. Simulation is a kind of sophisticated dry run at a set of future circumstances. It is used to rehearse, practice and promote new skills or ways of working at individual, team or system levels. According to what Patterson, Deutsch and Jacobson write in Chapter 13, team- and systems-level simulation point to a way to better understand resilience. They argue that *in situ* simulation, orchestrated by front-line carers on front-line problems, can help overcome the WAI–WAD difference by rehearsing how clinical groups do things in realistic environments. This performance can be contrasted with what clinicians are 'supposed' to do according to WAI, and what they later do, in practice. Simulation is used to experiment with new ways of doing things (running through a new procedure say, or a novel treatment mode which is being enabled in conjunction with a new piece of equipment) and in the process bring to the surface shortcomings in the landscape of extant guidelines, procedures or resource allocations. This performance can then be adjusted, or the new behaviours reconciled, with the WAI landscape. The preparation that simulation encompasses might in turn increase capacity and margins to manoeuvre down the track when the inevitable perturbations, or even crises, emerge. In this regard, simulations can bring to attention not just the core and range of clinical activities, but also the boundaries of systems performance. Patterson, Deutsch and Jacobson observe that simulation goes beyond imagining how a system performs, and tests its functionality. It enables that knowledge to feed into anticipating future capacities and strengthening its future manoeuvrability.

Modelling and simulation might indeed be of help to those hoping to explore and ultimately reconcile WAI and WAD, but what about the contribution of a more direct method – training? All too often in health care, training programmes are introduced as a unitary solution in a system that is expected to be static. Under this kind of regime, people discern they have a problem, or deficit, and then decide the remedy is to train the staff. But training is not just training and it is never done in a stationary, unidimensional system. Training is a complex phenomenon. It is an intervention that promotes learning and by its very nature enacts new, iterative relationships *in situ* between richly interacting stakeholders. Against this background, Clay-Williams and Braithwaite in Chapter 14 explore the what, who, when, why and how of training in complex health settings. They conclude that we need to exercise caution in imagining that training is a simplistic solution, and we also need to seek sustainability in the dynamic changes that training provides. This means, *inter alia*, evaluating progress and adjusting performance over time in response to training.

In a final chapter under the Methods and Solutions heading, Wears and Hunte in Chapter 15 look at the actions people take in designing and then instituting procedures. By procedures they mean the full gamut of WAI-based guidance tools including formalised rules, policies, checklists and guidelines. These solutions are typically drawn up and then mandated or rolled out. Although procedures mostly seem constraining rather than liberating, Wears and Hunte note that this need not be so. Procedures can represent accumulated wisdom, and be resources for novices and guiding templates for more experienced practitioners. Procedures can provide background information about what a collective holds to be common ground and be the reference against which adjustments can be consciously made when the need arises. Procedures can also act as a blueprint for slavish adoption. Wears and Hunte conclude by making suggestions for creating better procedures: ones that could contribute to rather than diminish resilience. These include adding a multiplicity of viewpoints into the development of procedures; explicitly looking at the differences between WAI and WAD; keeping procedures on top of changes in skill mix, technology and capacities; and monitoring the contributions procedures make to care delivery over time.

In short, procedures and their cousins – policies, regulations, checklists and guidelines – should ideally be made to be flexible, and not be overly detailed. Over-specifying always runs the risk of over-constraining people. 'Pretty good' procedures are likely to outperform over-specified ones if, as is the case in health care, there are well-trained, knowledgeable professionals with high levels of expertise and commitment on the front line. The bottom line is that we want to promote adaptive behaviours though intelligent, flexible procedures that suppose that coalface people are intelligent, flexible operators. The converse, dumb, rigid procedures are predicated on people who are dumb and rigid. In health care, people on the front lines are almost never that. So why insist, ask Wears and Hunte, on procedures that imagine they are?

12

Modelling Resilience and Researching the Gap between Work-as-Imagined and Work-as-Done

Janet E. Anderson, Alastair J. Ross and Peter Jaye

CONTENTS

Introduction

One of the key insights of resilience engineering (RE) is that work-as-imagined (WAI) and work-as-done (WAD) are different (Woods and Hollnagel, 2006). Assumptions about how work is accomplished are often very different from the reality of everyday clinical work and the actions that workers have to take to achieve the goals of safe, high-quality health care (see e.g., Debono and Braithwaite, 2015). Assumptions about how work should proceed are pervasive; they are enshrined in procedures and protocols, and are implicit in attempts to improve health care quality by introducing quality targets such as the requirement to treat and discharge patients within a certain time. Processes for investigating adverse incidents are often based on assumptions that (1) incidents can only arise from work that is not safe, (2) what constitutes safe work has been fully specified in advance and (3) therefore the causes of incidents can be identified by simply examining how people's actions deviated from these written specifications. Health care has been characterised as a complex, adaptive system (Braithwaite et al., 2013a; Robson, 2013, 2015) meaning that linear tools and methods are limited in what can be achieved because they provide limited understanding of how outcomes are achieved.

Assumptions about WAI often turn out to be wrong because they are based on a fundamental misunderstanding of the health care environment.

It has been pointed out that, in a dynamic health care system, extant rules are rarer than is often thought, and the 'area of action covered directly by clearly enunciated rules is really very small' (Strauss et al., 1963, p. 189). Many studies of technical work in health care have shown that the path to success for clinicians lies in finding a way through situations characterised, not by easy-to-follow guides, but by many competing priorities, conflicting demands and organisational shortcomings that have to be adapted to and overcome to ensure a good outcome for patients (e.g., Roberts et al., 2005; Nemeth et al., 2007, 2008; Dekker et al., 2013; Sujan et al., 2015b). The learning from these studies is that clinicians' ability to successfully navigate this difficult landscape creates safety and good outcomes for patients, in direct contrast to the view that safety arises only from following what has been pre-specified in rules and procedures.

RE argues that clinicians create good outcomes by adapting appropriately to the demands they face, but adds that their attempts to deal with the demands and problems of the workplace can also create negative outcomes (Dekker et al., 2011; Patterson et al., 2006a). In RE terms, bad outcomes are seen as aetiologically equivalent to good outcomes because they both arise from the features of the work system that require adaptive behaviour by clinicians. Although outcomes may be good or bad it may not be possible to describe the actions that preceded them in those terms.

RE is based on a systems perspective and, through this lens, health care is characterised as a complex adaptive system: 'complex in that there are a large number of interacting ... services, agents and processes, and adaptive in that the system is able to self-organise and learn' (Clay-Williams, 2013, p. 124). According to Wallace and Ross (2006, p. 170), 'Complex or non-linear systems are characterised by interconnected subsystems, feedback (and feedforward) loops, multiple and interacting control systems, and an incomplete understanding ... regarding how the system really works'.

Clinical work will always require decisions about how success is defined for each patient, how to manage competing demands, which goals to meet first, which of the required resources are unavailable and how to manage despite this. And these decisions will often be based on incomplete understanding, because, for example, 'agents in specific localities are unable to apprehend all behaviours of those in other parts of large systems ...' (Braithwaite et al., 2013a, p. 59). It is certainly necessary to plan work in advance, but this WAI ('mapped out') by those removed from the sharp end is not the same as work as it is achieved in practice not least because the plan fails to recognise how people will necessarily and actively *create* safety despite the tensions and contradictions encountered (Sujan et al., 2015c). It is not possible to fully specify in advance how work will be accomplished because of the dynamic, social and unpredictable nature of the environment (Patterson et al., 2006b). This gap between WAI and WAD is a key feature of many complex work environments but is nevertheless unacknowledged in most patient safety discussions.

These are powerful insights that are in direct conflict with many of the assumptions and accepted practices of quality improvement, rooted as they are in biomedical models of disease and illness that imply that quality problems can be diagnosed and remediated by targeted treatments. It is less clear, however, how these insights can be harnessed to inform quality improvement efforts. For example, how can we move from recognising the gap between WAI and WAD to studying it and understanding it? How can understanding the gap inform quality improvement? If good and bad outcomes emerge from the same features of health care work how can we target our improvement efforts? How can we learn from what goes right when good outcomes are overwhelmingly more common than bad outcomes and with the limited resources researchers have to focus their efforts? How can the field move beyond descriptive studies illuminating the dilemmas, workarounds and trade-offs of clinical work to address how to improve quality?

Redefining the Problem

If the gap between WAI and WAD is ubiquitous, why should it be a focus of attempts to understand how to improve quality? Imagined work is embodied in targets and procedures originating from organisational management, regulatory bodies and policy makers, but is not always based on a realistic view of the demands of clinical work. Clinicians are aware that if there is an adverse event someone who has deviated from procedures will probably be blamed, even though others may have taken the same actions many times before with no adverse consequences, the procedure may have been less than fully clear at the time, or a mindful procedural 'deviation' may have been deemed necessary to ensure the delivery of care. For managers, the gap between WAI and WAD means that their mental model of how work is accomplished is not realistic and they risk making decisions that stifle clinical work if they do not involve their staff in decisions that affect the flow of work. They may also not be cognisant of any new risks that might be introduced by workers' adaptations and workarounds and may not be able to intervene to ensure that adaptations are appropriate. One simple approach seems to be to make WAI more like WAD, that is, to better specify and more closely align what the rules say to how work is actually played out. This alignment may be possible to an extent, but in this chapter we argue that aligning WAI and WAD completely is not possible, and further that the gap should be acknowledged and managed by supporting safe adaptation to the varying conditions of clinical work. We therefore suggest that the focus should be on understanding the gap between WAI and WAD.

If RE is to contribute to improving the quality of health care – and we believe that it should and can – there is a need to research the gap between

WAI and WAD and understand how it drives behaviour and outcomes. Only by understanding the links between these can we gain insights into how to harness RE ideas to improve care. Others have also made these points (e.g., Patterson et al., 2006b; Nemeth et al., 2008; Dekker 2003). We want to reiterate this point and suggest a way to facilitate progress. Our view is that the problem to be addressed is not how to realign WAI and WAD (which we will argue cannot be achieved) but to understand how to make it a tractable problem amenable to research. Specific research questions that have not been addressed to date include

- What conditions in the work environment create the need for adaptive behaviour (WAD)?
- What pressures or problems are clinicians responding to when they create new ways to achieve outcomes (WAD)?
- Do new ways of working (WAD) not enshrined in policies and procedures (WAI) eventually become codified in practice? If so, what drives this process? If not, why not and what are the consequences?
- How and under what circumstances do adaptations lead to successful and unsuccessful outcomes?

Answering these questions will enable us to understand better what combinations of work system conditions and adaptive behaviours lead to what outcomes and provide insight into how we might begin to devise ways to strengthen the pathway to success and dampen the pathway to failure. A working model of these ideas will help to clarify this.

A Model of Organisational Resilience

The need to model resilience was identified some years ago by Hollnagel and Woods (2006) who argued that a successful model helps to focus attention on important concepts or aspects of a problem, and to understand mechanisms by explaining relationships between the concepts. Figure 12.1 shows our working model of organisational resilience that we are using to guide ongoing empirical work.

Like all models, it does not show all the detail in the real world; it is an abstraction that allows us to identify the theoretical concepts and their relationships that we are interested in investigating. In this model WAI is conceptualised as the interplay of two elements – demand and capacity (pictured on the left of the model). Demand refers to pressure in the clinical environment and includes requirements for effective care, such as the targets and standards set by regulators and policy makers. The nature of clinical demands

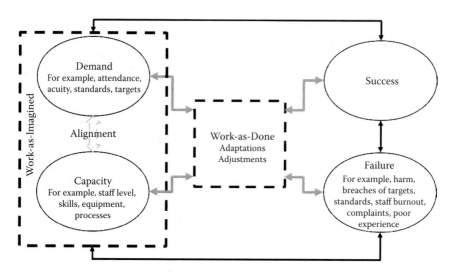

FIGURE 12.1
Working model of organisational resilience.

varies depending on the clinical speciality. For example, in maternity care the patient characteristics, their clinical needs and their acuity are very different from those in surgery, creating different demands. Nevertheless there will be care to deliver, and this demand is met by a range of capacities, including numbers of staff, their skill mix, physical infrastructure and equipment, processes, procedures and protocols. Note how processes, procedures and protocols in this model are simply aspects of capacity that interact with other capacities, such as staff resources and equipment availability. Processes, procedures and protocols are viewed as tools to be used by clinicians to meet multiple demands rather than prescriptions for action.

The model proposes that a presumed alignment between demand and capacity represents WAI, captured in the model by the dotted box surrounding demand and capacity. Organisations strive to align demand and capacity. Decisions about alignment are based on how work is imagined and described in procedures and anticipated demand based on prior experience. In the imagined world, either capacity is matched to demand, or potential shortfalls are identified and contingency plans are developed to address these. Despite these efforts, we contend that demand and capacity are fated to be always misaligned. The clinical demands and the work environment are too complex, and there are too many possible demand–capacity interactions, for the two to ever match completely. Thus WAD is characterised chiefly by dynamic adjustment to the multiple ways in which demand/capacity can be misaligned and the emergent effects of these complex interactions.

Factors such as staffing or equipment unavailability, a less-than-optimal skill mix or an unforeseen surge in patient numbers will be predicted *to a certain extent* and a plan for alignment drawn up, in the short, medium or longer term. However, clinicians will inevitably face challenges that have not been and cannot be foreseen. Adaptations and adjustments are thus a central feature of clinical work and are shown in the centre of the model. Adaptations are driven by mismatches of demand and capacity and require clinicians to work around problems and devise solutions, such as rearranging the way that care is delivered so that it is more efficient. Formal contingencies put in place simply cannot allow for all issues that will emerge from the interplay of factors such as patient acuities and comorbidities, staff skills and experience, equipment availability and functioning and policy and procedural conflicts. Further, it has been pointed out that the required trade-offs that attempt to resolve tensions in themselves create tensions of their own (Sujan et al., 2015c).

The extent to which organisations are *mindful* of these adaptations (i.e., they are openly discussed, monitored and understood) is likely to be an important aspect of success. Adaptations made *in situ* as work is experienced may be incorporated into formal procedures and become codified as organisational responses to the misalignment of demand and capacity. For example, adjustments made to flexibly manage mismatches between demand and capacity in a busy emergency department may eventually be formalised in an escalation policy where inbuilt triggers or alerts lead to reconfiguration of resources (capacity alignment) when demand is high. The important thing to realise is that the process conceptualised in Figure 12.1 is recursive. The incorporated policy now constitutes *the new imagined world*. There will soon be new misalignments, new complexities, new difficulties to deal with and new adjustments to monitor and understand.

Outcomes in the model are linked to adaptations introduced to deal with misaligned demand and capacity. Because adaptations vary in their effectiveness for dealing with pressures and demands, outcomes also may vary. In some cases adaptations can lead to adverse incidents. For example, the need to work more efficiently may result in gaps in documentation and a patient receiving the wrong medication. Adaptations thus mediate between 'demand–capacity' and 'outcomes', as shown in Figure 12.1.

In the recursive model, experienced outcomes, both successful and adverse, affect perceptions of demand, capacity planning and adaptive behaviour and these influence ongoing performance. Changes in any of the elements of the model affect other elements. Successful outcomes such as timely treatment and discharge may reduce demand by reducing readmissions, whereas unsuccessful outcomes may increase demand through the imposition of targets or lead to increases in capacity via training and skill development. Feedback can also occur between outcomes and adaptations – people's experiences of success may lead to overconfidence in improvising procedures whereas failure may lead to reluctance to improvise and outcomes such as patient experience may suffer due to, for example, longer waiting times.

In summary, we hypothesise, based on RE principles, that adaptations, driven by the misalignment between demand and capacity, lead to both success and failure. Guided by the model we intend to test these relationships empirically. These hypothesised interactions between the elements of the model are amenable to empirical investigation and the findings will help to identify opportunities for improving quality. Based on preliminary work, we can propose two possible avenues for improvement:

1. Closer alignment between demand and capacity will potentially reduce the requirement for adaptations *in situ*. Although the effects of this are difficult to predict we assume that reducing the need for adaptations will lead to better planned, coordinated and controlled adaptations which in turn may reduce adverse outcomes. We do not propose that it is possible or desirable to eliminate *in situ* adaptations. Rather, we envisage that reducing the need to adapt will create a more stable environment, reduce clinician stress, reduce the cognitive load imposed by the need for adaptations and lead to those adaptations that cannot be avoided being better planned. In turn, adverse outcomes may be reduced.

2. Improving adaptations is a key to improving quality. This can be done by strengthening successful adaptations and dampening those that are risky or unsuccessful. This will require a deep understanding of the adaptations in a particular clinical setting, what drives them and how they lead to success or failure. These hypothesised mechanisms can be investigated empirically.

Researching Resilience

Researching these subtle and changing phenomena and identifying how the interplay of environmental, organisational and behavioural factors leads to outcomes is challenging. Although resilience has positive connotations and implies that successful adaptation has been achieved, exhortations to learn from 'what goes right' are rarely helpful because researchers gathering data in the field need a more defined focus than simply everything that goes right. The model described in the previous section has helped to refine our research questions and focus; we are interested in the misalignments between demand and capacity that lead to the need for adaptations, and in understanding how clinicians succeed despite pressures that might, under other circumstances lead to failure. In other words, we are interested specifically in learning how clinicians succeed when under pressure (misaligned demand–capacity) and there is a clear risk of failure. We propose that these questions can be approached both qualitatively and quantitatively and that both approaches are needed to provide a complete understanding of how to strengthen care quality based on RE principles.

The need for in-depth qualitative methods such as ethnographic observations and interviews, participant and non-participant observations, case studies and field studies to understand work in complex adaptive systems has been recognised by others (Hollnagel, 2013; Robson, 2015). In qualitative work our aim is to describe in narrative the link between the main concepts in the model: WAI, WAD and outcomes. This requires an in-depth ethnographic approach to understand the complexities of the work environment, how it varies and how work is achieved. Such an understanding can only be built up over time in the field and with good rapport with clinical partners. Over time the main sources of pressure in the work system can be identified and the varying ways that clinicians deal with these pressures can be observed. Outcomes can be identified via observation, analysis of incident reports or discussion with the clinicians. Outcomes may include adverse events, but understanding outcomes requires a more subtle approach than focusing only on counting occurrences or identifying root causes by focusing on linear cause-and-effect relationships. The concept of non-linearity in health care has become somewhat fashionable, despite a lack of empirical demonstration that its mathematical properties are displayed. Nevertheless its clear rhetorical implication for applied work in health care is that rational reductionism (attempts to understand or improve the system by focusing on one part at a time) will not work (Wallace and Ross, 2006). As in many complex environments, there are multiple interacting goals and potential outcomes that may be of interest, such as patient experience, safety, efficiency, etc. These goals will conflict; clinicians will not be able to meet all goals when time and resources are limited. Understanding how goal conflicts are resolved and which outcomes are prioritised under what circumstances is an important focus; improvement initiatives need to embrace and understand this dynamic environment, rather than assume that a linear structure can be imposed and that such conflicts can somehow be ironed out.

These phenomena can also be investigated quantitatively and modelled statistically. Demand and capacity are quantifiable using direct and/or proxy measures. Staff numbers, skill mix, patient numbers and acuity are important, but there are many more demand and capacity variables such as equipment availability, beds available, discharges and ambulance arrivals that could be predictive of outcomes. *In situ* adaptations such as reconfigurations of staffing and resources, reallocation of staff and local adaptations to procedures can also be quantified. Data on outcomes are readily available, including adverse events, length of stay, target breaches, patient satisfaction and complaints. Modelling the relationships between these variables statistically is possible but challenging, largely because of difficulties with data capture and data quality. However, it has the potential to provide new insights into how outcomes emerge from the complex interplay of demand, capacity and adaptations.

Conclusions

In this chapter we have presented a theoretical model for investigating resilience in health care. The aim of our work is to identify potential avenues for intervening to improve the quality of care through RE principles. We have argued that this requires an understanding of the types of pressures encountered in the clinical work environment, what causes them, how they drive adaptive processes and how the interplay of these leads to outcomes both good and bad. Our thesis is that the gap between WAI and WAD cannot be closed because there will always be unforeseen circumstances and because clinicians have the autonomy and expertise to devise novel practices to achieve success. We have identified the misalignment between demand and capacity as a useful framework for investigating pressures and the adaptations that follow, successful and unsuccessful. These can be studied using in-depth qualitative methods for probing how work is achieved despite pressures, and quantitative modelling. Opportunities for intervention are likely to focus on improving alignment *per se*, being mindful of the adaptations and adjustments that will still always be necessary, and on strengthening successful adaptations and dampening unsuccessful adaptations.

13

Simulation: Closing the Gap between Work-as-Imagined and Work-as-Done

Mary Patterson, Ellen S. Deutsch and Lisa Jacobson

CONTENTS

Background

Simulation can be used as a tool in the quest to better understand and improve resilience. The term 'simulation' often conjures up a specific application and context of simulation, typically addressing individual or team skills or behaviours, but the field of simulation encompasses a broader array of capabilities with an impressive breadth of possible applications. Simulations can be summative, directed towards an established correct or best practice and designed to determine whether the participant(s) measure up or cross some threshold of competence or mastery related to pre-determined criteria. Simulations can also be formative, directed at exploration and learning. In a formative simulation, there may be a general objective, such as evaluating whether a new patient care space is suitable and ready to be opened for patient care, but there may not be an *a priori* 'best' answer. Either process (formative or summative) can be appropriate in different circumstances, but relative to understanding resilience, the formative process could be particularly valuable.

Many people are familiar with the use of simulation to improve the skills of individuals. These skills may be technical, such as the ability to suture or insert an intravenous line or perform an operative procedure; or non-technical, such as the ability to obtain informed consent or deliver 'bad news'. In some ways this use of simulation fits with an understanding of work from a framework that is integrated 'vertically'. For example, the 'sharp-end' or 'front-line' provider attempts to comply with and integrate the components of care delivery that have been provided or constrained by decisions closer to the 'blunt' end of the process.

Many are also familiar with the use of simulation to improve team skills. Again, team skills may be technical, such as coordinating the technical skills of a number of individuals in order to accomplish a task that requires diverse skills (e.g., resuscitation, surgical procedures) or non-technical, such as using closed-loop communication, resolving conflicts or demonstrating leadership and followership. In parallel with the concept of vertical integration noted above, the teamwork perspective of work provides a perspective from a 'horizontal' framework.

There is a third application of simulation, in which simulation is used to improve, or better understand systems; this may be the use of simulation that provides the most support to our goals of understanding or enhancing resilience. In this case learning can occur at an organisational, as well as individual level. For example, a simulation scenario, a 'situation' that requires interactions and responses to the circumstances or condition of a simulated patient, can be used to allow health care providers to care for a 'patient' in a new patient care area before opening the area for real patients. Or simulation can be used to examine a patient transfer process, or develop a new patient transfer process.

In each of these applications of simulation, debriefing is essential. The goal in the debriefing of these simulations is to discuss and learn from the participants' experiences and insights. This learning may rely on information from an expert (e.g., a certain medication or process is known to be the best option in this circumstance) or the participants may themselves be the experts, seeking together to develop an optimum understanding or approach. The debriefing discussion could address workarounds that were performed and resources that were used or, conversely, resources that were desired but not available; these resources might be cognitive, technical, informational; related to equipment, supplies, protocols; or whatever else is appropriate.

The opportunities for improved understanding and better performance are richest when the simulations occur *in situ* (e.g., an actual patient care setting) using real teams, real equipment and real clinical spaces, and evaluating actual processes. Other ways to learn about work-as-done, such as group discussions, analysis of large data sets and simulations that take place in a simulation centre, can provide useful information. But participating in simulations *in situ* can provide unique value. Although there are some simulations that cannot be accomplished *in situ* for logistical, economic or ethical reasons, *in situ* simulations offer the ability to come as close as possible to real work-as-done.

Goal of Simulation Relative to Resilience Engineering

The argument has been made that one cannot just decide to be resilient, or just decree that a system must be resilient. The goal of participating in simulation scenarios is not to train health care providers to be more careful, to practice vigilance, to develop rote responses or to develop standards or processes that 'guarantee resilience'. The goal is to better understand the components of health care delivery (equipment, medications, processes, knowledge) that impact the care of the simulated patient, and to experiment with variability – perturbations – in a setting that comes closer to actual patient care than many other processes. Simulation provides the opportunity to practice and to develop expertise and insight, to improve fluency, to 'unburden' the participants from lower-level functions and enable the capacity to adjust to novel situations or complex, rapidly evolving circumstances. Simulation can 'make visible' inadequate resources, or the inability to perform the necessary work in a time-pressured environment in a reliable or desirable way. Some of these findings may come from participants, who are 'content' or 'subject matter' experts; some may come from outside observers who bring fresh perspectives and different theoretical frameworks to the process. Ideally both viewpoints are contributory. The simulation will only be successful in the long term if the organisation is able to respond to the learning and/or 'uncovering' that occurs during simulation.

To provide functional examples, one outcome of the simulation could be to change resource availability, such as changing the location of specific equipment so it is more accessible; re-writing an electronic order set so it reflects the way in which clinical work is actually performed; reassigning individual team member responsibilities based on observed workload, etc. At a higher level, the participants may gain awareness and increased understanding of the system they are working within and the many parts that contribute to that system. Participants may need to re-prioritise tasks and goals in real time during an evolving patient care situation. Ideally, participants are empowered and develop the capacity to make system changes. The following case study illustrates this concept.

Case Study 13.1

A 14-month-old child has a history of a tethered spinal cord, a congenital condition in which tissue attachments limit the movement of the spinal cord within the spinal column (American Association of Neurological Surgeons, 2014). Four days after the child undergoes an uncomplicated surgical procedure, he is admitted to the paediatric intensive care unit (PICU) with a high fever, low blood pressure, lethargy and redness and swelling at the surgical site. The child is in shock.

As the child arrives in the PICU, an emergency response team coalesces. The team leader has studied this sort of crisis, but has limited experience taking care of a real patient with this specific condition. The leader assigns roles to the other physicians, nurses, respiratory therapists, etc. The team rapidly evaluates the child's circulation, airway and breathing (CAB). Team members report vital signs and other physical exam findings to the team leader. The leader synthesises this information and clearly states a mental model of a patient with septic shock to the team. Emergency vascular access, an intraosseus line inserted in the tibia, provides a means to emergently provide medications and fluids. Appropriate fluids, antibiotics and vasoactive medications are provided. During the resuscitation, the nurse informs the team leader that the requested epinephrine dose is incorrect, and the dose is corrected and administered. The child's condition stabilises. The team has generally performed well, but there are opportunities for improvement that are discussed during the debriefing following the resuscitation.

Their patient, despite having palpable pulses, chest wall motion while breathing and vital signs displayed in real time on a monitor, and despite responding to the administration of medications and other interventions, is only a simulator.

The team then debriefs with a trained facilitator to discuss their reactions to the scenario, the teamwork and communication processes, the medical management, resource and system issues. In the follow-up a written summary will be developed that addresses

- The topic of the simulation, including learning objectives (perform a CAB evaluation; identify septic shock; implement correct treatment based on best evidence).
- The principles of teamwork and communication ('The team leader assigned roles quickly and elicited feedback from each team member.' 'Team members were willing and able to adjust roles depending on their abilities'.).
- The principles of medical management for this condition.
- The finding that the medication nurse's workload was noted to be overwhelming for one individual, negatively affecting the entire process and potentially requiring a reallocation of resources.
- Where to find supplementary educational information.

Clay-Williams and Braithwaite posit that training of health care providers in a complex adaptive system should incorporate the four cornerstones of resilient behaviour (see Chapter 14). Table 13.1 illustrates how this simulation scenario addresses the characteristics of resilience.

More importantly, when these team members encounter a 'real patient' presenting with septic shock and requiring resuscitation, the team's 'practice' with simulation and familiarity with the 'routine' management of a crisis, may create additional capacity in the team to deal with the rarely encountered clinical features of the individual patient. This idea of 'practice' aligns with the idea that "resilience is likely to be higher in organisations where risk is greater ... The more practice people get doing work or resolving crises, the better they become at it" (Clay-Williams and Braithwaite, 2013, vol. 1, p. 207).

TABLE 13.1

How Participation in the Simulation of Child Requiring Resuscitation Supports the Emergence of the Primary Activities of Resilience Engineering

Characteristics of Resilience	During the Medical Crisis	During the Debriefing	System Opportunities
Respond Know what to do, be capable of doing it; address the actual	Leader assigns roles Leader states a diagnosis which triggers patterns of response and interventions	Reinforce appropriate responses	Integrate this constructive debriefing process into actual patient care
Anticipate Find out, know what to expect Address the potential	Stating the diagnosis triggers accessing relevant knowledge	Elucidate patterns	Develop awareness of capability of team members
Learn Know what has happened Address the factual		Recognise pattern, provide references that are meaningful at that time, reinforce appropriate activities	Organisation may learn what resources were important, which were available, which need modification
Monitor Know what to look for Address the critical	The nurse recognised an incorrect dose, and provided that information to the leader	Discuss cues Encourage self-reflection	Develop/use early warning scoring system Improve ability to monitor for weak signals

Health care is inherently complex and high risk; however, the individual health care provider may experience a specific high risk event only infrequently. Simulation provides the opportunity for health care providers to develop and practice general approaches to problem solving, management and adaptive capacity during a clinical crisis.

Specifically, Sutcliffe and Weick have posited that "recognising, rescuing and managing emerging complications" are part of mindful organising. In health care, mindful organising:

> Provides a basis for people to interact as they develop, refine and update shared understandings of the situations they face and their capabilities to act on those understandings. When ... members focus sustained attention on operations (e.g. how their work is unfolding presently) they enhance the likelihood that they will develop, deepen, and update a shared understanding of their local work context and emerging vulnerabilities. (Sutcliffe and Weick, 2013)

Simulation has the potential to play an important role in this context. It is not possible to imagine every possible rare event and each health care provider is likely to have limited experience with a particular rare event. However, simulation experience provides an opportunity to practice the behaviours that enhance resilient performance during a crisis. Practicing 'noticing', 'communicating', 'updating' and 'refining' understanding during a simulated event creates a foundation and potentially increases the health care team members' situational awareness and adaptive capacity during an actual crisis.

Systems and Simulation in Resilience Engineering Terms

Expanding on the ideas described previously, we propose that simulation, when framed by resilience engineering, can be used as a tool to help us understand the conditions under which a system responds in a more or less adaptive fashion. On one level, simulation may allow us to see the unintended consequences of imposing constrictions on a system by creating an environment in which we are able to gather observational data. It allows us to see work-as-done (or very close to work-as-done). The utility of simulation in evaluating systems can be seen in multiple realms: pretesting a clinical space before an official opening using simulated patients but an otherwise realistic clinical environment (Geis et al., 2011; Kobayashi et al., 2006; Gardner et al., 2013), evaluating the preparations for a natural disaster using a mass casualty drill or even virtual worlds (Franc-Law et al., 2008; Ingrassia et al., 2012; Araz et al., 2012), even evaluating the usability of equipment before introducing it into a clinical arena or refining the process of new equipment implementation to minimize the hazards of exchanging equipment during actual patient care (Kushniruk and Borycki, 2014; Kobayashi et al., 2013; Anders et al., 2011; Geidl et al., 2011; Landman et al., 2014).

The use of simulation to evaluate reconfiguration of current emergency teams for a new emergency department provides a proof of concept. Simulations resulted in redefining the scope of practice for various resuscitation team members as well as the addition of a pharmacist by a remote link. The resuscitation bay was also reconfigured to provide separate areas for medication preparation in the event that two critical patients might require simultaneous care (Geis et al., 2011). This reconfiguration and redeployment of resources occurred in a relatively short time frame of approximately 6 weeks. The health care workers iteratively participated in simulations, debriefed, suggested solutions and tested solutions. This use of simulation allowed the team to improve their adaptive capacity and margin for manoeuvre proactively, even before actual patients were cared for in the new facility.

In these settings, we can observe cross-scale interactions and evaluate to what degree a system has performed according to expectations. Has the organisation's choice of a new point of care laboratory test facilitated faster turnaround times and therefore improved throughput? Or, were the work-arounds that the staff had already implemented to maintain throughput now in direct conflict with the new protocol? Did the organisation choose the right goal for improvement? Obviously the changes and disasters imagined are limited to our human creativity, but the ability to see how the system functions in response to what we can imagine may prevent us from engi-neering out those components that keep the system running. Simulation of new processes prior to implementation may help us avoid creating increas-ingly brittle systems. Conversely, when planned simulations are cancelled because of a lack of resources, questions arise as to whether that system is functioning with insufficient margin should another threat, such as addi-tional patients requiring care, be encountered.

Simulation may also promote freedom of development, unburdening the actors in a system from a brittle, rule-intense environment, and unburdening them from concerns about immediate harmful consequences for a real patient thus allowing them to create new solutions (aka workarounds). Medicine, like many fields, is full of innovations created when individuals lacked the equip-ment necessary to perform a task. This can be as simple as using tubing nor-mally used to deliver oxygen to the nose to instead irrigate a foreign body out of the eye with saline. It can also be significantly more impactful, such as family members wearing layers of plastic trash bags over their bodies while caring for loved ones stricken with Ebola. At times instead of being inno-vative because of an under-resourced setting, individuals instead develop workarounds to save time or money. These are most evident when increased constrictions are placed on an operator's resources (personnel, time and equipment) and there are increased expectations for productivity. A system may be under-adapted secondary to pressure and restrictions from stake-holders, which may be well intended but may contribute to maladaptation because of the resulting trade-offs. The Law of Stretched Systems holds that because of resource and performance pressures, we tend to take the benefits of change in the form of increased productivity and efficiency rather than in the form of a more resilient, robust and therefore safer system (Woods and Cook, 2002). Simulation can make these trade-offs explicit as well as demon-strating the effect of these changes on adaptive capacity of the system.

One significant limit to fluent behaviour is cognitive overload. When a sys-tem is already stressed by anticipated or known threats and then encounters unanticipated threats, it is optimal that existing adaptive capacity would prevent failure and at least facilitate graceful degradation if not usual opera-tions. A risk to operating in an uncontrolled environment is overburden-ing the individual with extraneous cognitive load. Simulation activities are a useful mode of training for these outlier situations, allowing at times for the movement of tasks from intrinsic to germane load. Simulation and psychology research suggest that stress (such as the need to operate in an overburdened, under-resourced environment) can increase cognitive load. (Some would posit that in certain health care environments, such as urban emergency departments, these situations are more the norm than the out-lier.) This can inhibit performance in novices, but may enhance performance in experts. We believe that by using simulation to develop expertise, stress related to cognitive overload becomes less of a factor, and the individuals and the systems in which they work may demonstrate graceful extensibility and adaptive capacity (Harvey et al., 2012; Schwabe et al., 2010; Fraser et al., 2012; Demaria et al., 2010; LeBlanc et al., 2008; LeBlanc and Bandiera, 2007; Driskell et al., 2001; Woods, 2014; Finkel, 2011). Adaptive capacity exists before stress occurs. Ideally, we can assess components of our adaptive capacity through exercises in the anticipation and reaction to past or anticipated disturbances. This falls within the purview of basic simulation assessment. We can stretch our assessment further by simulating imagined threats.

We propose an even larger role for simulation in the realm of work-as-imagined bringing us closer to work-as-done. Participants in simulation activities learn affective lessons (consciously or not) from their simu-lated experience, much like they do from their real clinical experiences. By modelling exploration and collaboration, we believe that simulation experiences cultivate desirable behaviours. By encouraging attempts to anticipate, monitor, respond and learn, we anticipate helping to build these skills. Further, it is critical that simulation training reflects work-as-done rather than work-as-imagined. To train for processes or situations that do not reflect actual clinical processes is frustrating and potentially harmful (see Chapter 14).

Paradoxically, simulation may also increase brittleness by making work-arounds visible and therefore subject to sanction and attempts to reinsert the standard work practice. This practice should not be the goal of such exercises and facilitators of the process should strive to reorient their priorities away from maintaining static gold standards and best practices and towards free-dom of exploration and ingenuity. The goal should be to create an environment for evaluation of the system, not the individual; or perhaps, an environment without evaluation at all. Rather, simulation may be viewed as a means to create an environment that promotes freedom to explore and experiment in order to avoid brittleness. This type of exploration reveals to participants what is actually occurring and the current capacity of the system. By allow-ing experimentation and problem solving within the simulated environment, operators learn how the system functions and responds to stress and pertur-bations and begin to understand the tolerance and margins of manoeuvre.

Resilience Engineering and Simulation in Practice

In a number of ways, some ongoing simulation work occurs with a resil-ience engineering perspective, albeit without articulating that conceptual framework. *In situ* simulation, simulation that occurs in the clinical environment (Patterson et al., 2008, 2013), is now frequently utilised as a tool to evaluate, identify and anticipate clinical risk for new facilities, teams and processes. This speaks to a willingness to anticipate and learn and is especially valued for high-risk clinical settings or processes such as emergency departments, labour and delivery, medical resuscitations and difficult airway response teams (Geis et al., 2011; Patterson et al., 2013; Wheeler et al., 2013; Johnson et al., 2012; Miller et al., 2008; Riley et al., 2010). In some organisations, *in situ* simulation is viewed as a necessary and expected element of the process of bringing new facilities online or prior to the implementation of new clinical processes.

Using *in situ* simulation to identify latent threats is only part of the process; organizations must also address the identified threats. Unfortunately, *in situ* simulations often have as their primary metric, identification of latent threats. Frequently, these latent threats are identified with respect to missing or malfunctioning equipment or supplies, or communication issues within the system. Although these are important, there are several concerns associated with a primary focus on latent threats from a resilience engineering perspective.

The first issue is that while latent threats are frequently identified, the methods by which organisations address potential risks vary widely. Frequently, there is no identified process for prioritising or resolving potential safety risks for patients (Auerbach et al., 2015). The paradox is that though the organisation typically supports the resources needed and the conduct of *in situ* simulation, frequently there is no process or mechanism to deal with the outcomes of *in situ* simulation. Thus, though risk is identified and understood by some members or micro-systems within the organisation, there is often no systemic response to the identified, but as yet unrealised threat. In fact, this represents a failure of learning from an incident, albeit a simulated incident.

Another possible consequence is that when *in situ* simulation is conducted, organisational leadership and management may believe that the potential threats have been identified and resolved for a particular situation. This premise suffers from two misconceptions. The first is that all potential threats can be identified using *in situ* simulation. The conduct of simulation is of course limited by the experience and imagination of those designing the scenario. Although individuals with experience and knowledge of the system have the capacity to identify regular and irregular threats (Epstein, 2008; Westrum, 2006) by definition, they will not have an ability to imagine the unexampled threat nor conceive of the fundamental surprise (Lanir, 1983).

In addition, it does not necessarily follow that identification of latent threats results in resolution of latent threats. Organisational leadership may not fully understand the threats that are identified and they may not understand the potential severity of the threats identified in simulation. The extent to which threats are formally surfaced to organisational leadership varies widely

(Auerbach et al., 2015). This may reflect a fallacy of centrality that managers in central positions (of authority) presume they would know if something important was happening, and so if they are not aware of anything going on, then nothing important is happening (Whetten, 1995). The danger then is that leadership is falsely reassured that the system is being tested and evaluated, and that the threats to safety and optimal system functioning are being addressed. This creates a miscalibration between leadership's perception of the system's functioning and safety and the actual performance of the system when stressed, even simulated stress.

Conclusion

In the best use of simulation in a resilience engineering framework, simulation becomes a means to understand the response of a system to various stressors as well as the adaptive capacity and margin of manoeuvre of the system. By allowing a simulation to run freely with skilled observation and facilitation, it becomes possible to arrive at an improved understanding of the workload of the team and the adaptations that are employed in order to maintain the functioning of the system. Debriefing allows the health care workers to reflect and surface perceived clinical risks as well as to suggest solutions. Simulation allows us to approach, that is, to come closer to work-as-done. Graceful extensibility of the system is identified and the extent to which health care providers adapt to stresses (which is sometimes invisible to system leadership) becomes apparent in a clinical simulation. In the best circumstances, simulation of processes in new or existing facilities can proactively identify the likely boundaries of safe system function. Rather than imagining how work is done, simulation allows us to anticipate and evaluate in a proactive way the adaptive capacity of the system in the face of unexpected challenges.

14

Realigning Work-as-Imagined and Work-as-Done: Can Training Help?

Robyn Clay-Williams and Jeffrey Braithwaite

CONTENTS

Introduction

In hospitals, training design and methods are frequently optimised for the linear system that many people imagine health care to be. Like other processes and procedures, however, training must be developed with adequate understanding of work-as-done, rather than work-as-imagined. Inappropriate training, poorly conceptualised training or training ineffectively applied can add an unwelcome layer of complexity to an already complicated health care system, with little discernible benefit. At best, this can be a poor use of resources. At worst, training can completely misfire, and cause serious harm.

This chapter discusses five aspects of health care training, using examples and lessons learned from a team-and-simulation-training study

(Clay-Williams et al., 2013): what do we need to train, whom do we need to train, when do we need to train, how do we need to train and under what circumstances do we know our training is successful?

What type of training is likely to help enhance naturally occurring organisational resilience? The linear 'organisation-as-imagined' perspective, for example, encourages aligning training with professional needs or sectional interests rather than those of inter-professional patient care. Imparting contextual knowledge on how the health care system actually functions might therefore be an early gambit worth supporting. We also need to consider whom we train. We could adopt a wider view of who are team members, and include leaders, patients and carers, along with clinicians; this could help realise a goal of enhancing the 'organisational mental model'.

The method of training may be as important as the content – practice in working through high-pressure scenarios as a team, via patient simulation or virtual reality modes, appears to show promise (Graafland et al., 2012; Ghanbarzadeh et al., 2014; Hansen, 2008). Evaluation of training and feeding back what is learned into improving training are also well-documented initiatives. Health care often treats training as a static system, whereby once the course has been developed the job is done (Weaver et al., 2014). In a resilient system, training is flexible in meeting the needs of learners as they evolve and adapt.

We know that health care is best characterised as a complex adaptive system (Braithwaite et al., 2013a). Although such a system will tend to self-organise for effectiveness, we can hamper that process by introducing ways of working that are incompatible with the way complexity manifests itself. Our ways of leading, management and training in health care are all too often designed for a hierarchical linear system, and we may need to rethink how we accomplish these activities if we want to enhance resilience. In a complex adaptive system (CAS), learning occurs as the result of iterative relationships between agents (Lichtenstein et al., 2006; Uhl-Bien et al., 2007), and is emergent. Therefore, our primary goal as a trainer is to facilitate productive interactions between clinicians, leaders and patients and their carers.

There is a large overlap between the concepts normally ascribed to teamwork and the functioning of a CAS. Teamwork is broadly conceptualised as the interaction between two or more individuals working towards a common goal (Baker et al., 2010). This is the same type of interaction where we might expect learning to occur in a CAS, although the degree to which teamwork will be effective will differ across any complex organisation.

What Do We Need to Train?

At bedrock, training is a method of education that provides a conduit for imparting or eliciting knowledge coupled with an opportunity to develop skills by translating that knowledge into practice (Goldstein and Ford, 2002). We can consider what we might need to train in terms of the knowledge, skills and attitudes that have potential to enhance system resilience. If we start with the four essential capabilities of resilience (Hollnagel, 2009b) – namely, knowing what to do, what to look for, what to expect and what has happened – we can narrow down what we need from training to having some degree of awareness of what is happening or likely to happen, being able to act to address an impending problem and being able to learn by self-reflecting on an event.

Knowledge

There are two aspects to knowledge training that may be useful to consider. The first is awareness training: what is this 'system' in which work occurs and tasks are enabled, and how does it really work? Health care professionals will act in accordance with their mental model of how the health care system works. Despite easy availability of evidence to the contrary, clinicians are continually bombarded with information that suggests that the system is linear: 'wiring diagrams' of the organisation show an hierarchical system, regulatory and quality frameworks (often based on ideas imported unquestioned from other industries) attempt to 'control' the system in a linear manner, the division of hospitals into management departments implies that the subsystems are bounded, and comprised of independent elements, and the categorisation of patients in terms of organ malfunction separates complicated and chronic diseases into a series of discrete problems.

If we take a step back, however, we can see that the health care system is actually more complex (Braithwaite et al., 2013a) than this. It comprises interrelated causal loops (Wears, 2010) and networks (Creswick and Westbrook, 2006), and weakly and strongly coupled behaviours depending on circumstances and settings (Braithwaite, 2010). Understanding that these characteristics describe how the system actually works may alleviate a degree of frustration, with clinicians less likely to find themselves having constantly to fit square pegs into round holes. Acting in alignment with the system as it is, rather than the system as it is imagined, has the propensity to facilitate development of processes that are safer, more effective and more efficient. Therefore, the initial step in training should be awareness of health care as a CAS.

A second aspect of knowledge training is the knowledge underpinning skills that may be useful for operating effectively in a CAS. Awareness, action and reflection, both individual and as a team, align with the basic themes of crisis resource management (CRM)-based training. In health care, CRM training has been shown to improve knowledge, attitudes and behaviours associated with team skills (Buljac-Samardzic et al., 2009; Salas et al., 2008a), and exhibits potential as a turning point for improving resilience. Of the wide range of identified training competencies for team skills, the knowledge underpinning situational awareness, communication, task management and decision-making and leadership skills (Clay-Williams and Braithwaite, 2009; Alonso et al., 2006) is likely to be useful in improving resilience.

Skills

Mindfulness training, leading to increased awareness of the present, can minimise biases (Sibinga and Wu, 2010), and enable health care professionals to focus on the actual situation rather than what they would like it to be. In a small-scale study in radiation oncology, for example, mindfulness training delivered to clinicians with the goal of decreasing errors resulted in increased reporting of near-misses (Mumber, 2014), indicating potential for this training to contribute to the monitoring capability of resilience. Situational awareness, or the ability to gather and interpret information about what is happening external to the perceiver, and to use that knowledge to predict future events (Flin et al., 2008) is an important skill for functioning in an uncertain and changing environment. Situational awareness is a component of teamwork training that has great potential in addressing the resilience capabilities of monitoring, responding and anticipating.

In addition to mindfulness and situational awareness training, there are other candidate skills to help. Team training addresses two types of skills – portable generic skills, and those specific to role. Portable skills can be learned by an individual and used as required to meet the needs of any team situation. Examples include communication skills, assertiveness training and conflict management techniques. Skills specific to a role are those, such as cross-training or cross-monitoring, that are relevant to specific roles or tasks. Examples include the team coordination aspects of resuscitation skills, and ISBAR (Identify, Situation, Background, Assessment, Recommendation of patient status) (Thompson et al., 2011), a standardised communication tool specific to patient hand-off.

With regard to the resilience capabilities of taking action or responding, teamwork training provides instruction and practice in task management, decision-making and leadership. In regard to individual and group reflection, teamwork training teaches skills in stress recognition, debriefing, feedback and asking questions (Clay-Williams and Braithwaite, 2009).

Attitudes

Attitude structure has been found to be important for learning (Warr and Bunce, 1995); however, attitudes can also be modified by learning team skills (O'Connor et al., 2008; Salas et al., 2008b). A prior positive attitude to a systems approach to safety would be useful; therefore, it is still possible to acquire or model this attitude as a result of training.

Whom Do We Need to Train?

The simple answer is that we need to train everyone, but of course in health care that is not so simple. Training in health care is frequently arranged along professional lines, and usually in accordance with the requirements for technical skills training. Hence, surgeons are trained in colleges of surgery, and psychiatrists in colleges of psychiatry, and so on. This cuts against the grain of cross-disciplinary team working. Multidisciplinary teamwork and inter-professional training are becoming more prevalent, but are not universal and do not normally include organisational leaders or patients. Mandatory training, such as occupational health and safety, fire safety, infection control, equity and diversity, can be implemented system-wide, but this is normally rudimentary knowledge training that can be, and increasingly is, conducted online.

Leaders

Where do leaders fit into the training plan? If the leadership team does not understand how work is done, and tries to apply top-down policies in accordance with work-as-imagined, the need for workarounds (and consequently the workload) will increase (Debono et al., 2013). This can lead to increased system complexity and propensity for error. In a CAS, leadership can be defined as an adaptive process that is dependent on context and that "emerges in the interactive spaces between people and ideas" (Lichtenstein et al., 2006, p. 2). It is important that health care leaders understand that it is not their role to 'act on' the organisation or to assert control, but rather to facilitate communication and collaboration within the organisation to ensure that effective processes that emerge are supported and sustained. In this sense, the leader is an enabler, and leadership becomes a shared construction: less heroic, individualised leadership and more distributed leadership (Greenfield et al., 2009). Providing training to facilitate the leaders' understanding of this process is likely to lead to better outcomes, and a less stressful work environment for all.

Front-Line Personnel

When considering whom to train, it is important to include the front-line doctors, nurses and allied health professionals, and also those who work farther from the coalface. This might include supervisors, professionals involved in clinical support roles such as pathology, radiology and pharmacy, and also ancillary staff such as wards persons, cleaning staff and administrators.

Patients

How do we include patients in the training? And their carers? Although patient-centred care involves patient participation and involvement as a core element (Kitson et al., 2013), this level of inclusiveness rarely extends to provision of training. Although clinicians will often include real or simulated patients in their own medical training, it is important to also consider training the patient (Longtin et al., 2010). Various ways of improving patient involvement in their own care have been investigated, including encouraging greater communication between patients and clinicians (Jangland et al., 2012), and improved patient participation in decision-making (Tariman et al., 2010; Wiley et al., 2014). In general, improved patient participation is associated with reduced adverse events, and with patient perceptions of better care (Weingart et al., 2011). In an effort to improve participation, patients are increasingly being provided with information about their rights and responsibilities. The World Health Organization has endorsed patient participation as a method of improving patient safety (World Health Organization, 2008), as have government and private health care organisations in the United States (American Hospital Association, 2003), United Kingdom (National Health Service, 2013), Europe (European Commission, 2002) and Australia (Australian Commission on Safety and Quality in Health Care, 2008). It is, perhaps, time to consider the patient as an integral part of the system when designing training, rather than an adjunct or beneficiary. Studies have found that patient education is a key predictor of participation (Longtin et al., 2010). The patient is a key source of the variability encountered in the workplace, and a patient who understands the nature of the health care system will be better placed to navigate that system and actively participate in his or her own care.

Trainers

Training in health care has traditionally been a ride-along aspect of patient care, and frequently occurs as an integral component of the care process. Trainers, therefore, are often not educational professionals, but experienced clinicians who teach both technical and non-technical skills as part of their normal work. Skills associated with facilitating, coaching and mentoring, however, are specialised skills that should be recognised as such and taught, in addition to awareness training on resilience principles and how the health care system works in practice. Train-the-trainer approaches to teaching have a place in health care, in terms of cost saving and increased credibility when using trainers who are already respected as experienced clinicians, and have been shown to be successful in teaching teamwork skills (Baker et al., 2003, 2005).

When Do We Need to Train?

For optimum success, training should be needs based; that is, training should be available as and if required to satisfy the needs of health care professionals. Specifically, however, there are a few junctures in a medical career when training is likely to offer more leverage for learners. Awareness training, for example, is best provided as a pre-vocational programme. Receptiveness to improving knowledge, skills and attitudes that may contribute to resilience is perhaps highest towards the end of undergraduate training, when students are curious about how the workplace they are about to enter functions. Awareness skills could be reinforced, via refresher training, after an adverse event or other system perturbation where health care professionals are challenged and need resources and support to regroup.

Training in basic team skills is typically included in most medical undergraduate programmes; depending on the content and training method, this may satisfy the requirement for initial resilience capability training. On entry to the profession, however, it is important to re-visit team skills from a multidisciplinary perspective, and opportunities should be provided to practice the skills with others in the workplace, ideally supported by coaching or mentoring (Weaver et al., 2010). Often, to synthesise skills effectively, simulation training is required, where clinicians can learn by trial and error in a low-stakes environment. This type of training is best conducted using the team that normally works together, as the problems each team will encounter in the workplace are likely to be context-specific. Recently, this type of training has also been shown to be effective for health care leaders (Cooper et al., 2011). An organisational training strategy might involve individual team skills training on entry to the workplace, reinforced regularly with multidisciplinary team training (via simulation or classroom training), and include focused discussion and group reflection after workplace significant events (Clay-Williams, 2010).

How Do We Need to Train?

It would seem counterproductive to teach how health care is a resilient complex adaptive system, but implement the training through a linear hierarchical programme. Therefore, we may need to rethink the traditional approach. We know that learning only occurs when the learner is receptive to the training, and the training is contextualised to match learner needs (Goldstein and Ford, 2002). So, do we make training available, or do we mandate training? In a recent study on teamwork training, we found that using volunteers who self-selected into the training programme had a flow-on effect, with learners sufficiently motivated to adapt the lessons for their own workplaces following the course (Clay-Williams and Braithwaite, 2014). Perhaps we should, in most cases, therefore make training 'available', and ensure easy access for health care professionals. This can be particularly effective for health care workers who work in shifts, or unpredictable hours, or work under variable patient demands where prior commitment to a training course some months ahead is not practicable. With current technology, there are many on-demand and flexible options for imparting knowledge, including pre-recorded webinars and podcasts, online training lectures and modular training packages (Clay-Williams et al., 2014).

Skills

Choosing a method for skills training is more complicated. Although individual training, such as mindfulness training, can be made available on demand, communication, leadership, decision-making and other team skills are only learned by practice and interaction with others. Although we may not wish to impose training, the need to train in groups in order for training to be effective means that training will need to be scheduled. Research on training using serious games (Graafland et al., 2012) and virtual worlds (Ghanbarzadeh et al., 2014; Hansen, 2008) indicates these methods have potential for supplementing or replacing face-to-face training sessions, although in most cases validation is required. Regular simulation training has been shown to be effective for practising teamwork, decision-making under pressure and load shedding (for a detailed discussion of this method, including its potential contribution to improving system resilience, see Chapter 13). CAS theory espouses that learning or adaptation often occurs as a result of conflict or tension (Lichtenstein et al., 2006), so we need to consider deliberately introducing stressors into practice scenarios. Although stress exposure training has been shown to be effective for escalation management in the military (Cannon-Bowers and Salas, 1998; Driskell et al., 2008) and other high-risk industries such as fire safety (Bergström et al., 2011), it has yet to be widely tested and implemented in health care.

Attitudes

Attitude training is perhaps most amenable to diffusion of innovation methods (Rogers and Shoemaker, 1971). In our team training research programme, we found that trainees felt the training was interesting and useful for their work (Clay-Williams et al., 2013), and were likely to mentor and educate others, and role-model the new behaviours when they returned to the workplace (Clay-Williams and Braithwaite, 2014). Attitudes can also be altered in team training by participation in case studies and guided discussion. We found, for example, that some participants began to understand the perspective of other clinical professions for the first time as a result of discussion activities in the multidisciplinary groups (Clay-Williams and Braithwaite, 2014).

Under What Circumstances Do We Know Our Training Is Successful?

Design and implementation of training is only a small part of the equation – we need to know to what extent the training is adopted, and ultimately, whether the training improves system resilience. Systematic methods for evaluating training, such as Kirkpatrick's framework (Kirkpatrick, 1978), are readily available, but are likely to only provide a proximal indication of training success. At present, we do not have scales or instruments to specifically measure system resilience. There are a number of validated scales to measure attitude and behavioural components of teamwork, such as the Safety Attitudes Questionnaire (Sexton et al., 2006) and the Mayo High Performance Teamwork Scale (Malec et al., 2007), that may be able to be used as interim measures while resilience scales are developed. The Resilience Analysis Grid (Hollnagel, 2011b) may also provide a measure of the capabilities of resilience (particularly if used to characterise the organisation both before and after training), but has yet to be articulated as a validated scale.

What we do know is that any result is likely to be heavily context dependent: it is important for the training to be flexible and adaptable to the needs of the participants on the day. This will make it very difficult to aggregate results in a meaningful way, and means that any evaluation is unlikely to provide evidence to support large-scale implementation of generalised training programmes. What training evaluation can provide, however, is a continuing source of timely information on the utility of the training in meeting the current needs of the participants. If thoughtfully applied, this feedback will allow the training to be adapted alongside changes in the health care system to enhance work-as-done. Finally, by making training available to all who are involved in the processes of health care – including managers, clinicians and patients – an organisational shared mental model will be developed that will help to reconcile the differences between work-as-imagined and work-as-done.

Conclusion

To design a training programme that helps align work-as-imagined and work-as-done, we need to decide *what* to train, *whom* to train, *when* to train and *how* to train. Table 14.1 summarises the main aspects to address when designing training in health care.

We also need to evaluate the training and ask not only 'has something been learned?' but also 'has the right thing been learned?' Finally, we need

TABLE 14.1

Considerations for Training Design

What?	Awareness Training	Team Training
Knowledge	Health care as a complex adaptive system	How to work more effectively with others in this system
Skills	Awareness skills, e.g., mindfulness training	Portable generic components of team skills training
Attitudes	Safety-I and Safety-II	Inter-professional teamwork
Whom?		
Leaders	Incorporate resilience principles into leadership training	
Clinicians	Train everyone – student clinicians, doctors, nurses, allied health professionals, ancillary staff	
Patients	Include patients and their carers	
When?	**Awareness Training**	**Team Training**
Knowledge	Pre-vocational	On entry to the workforce, reinforced regularly, e.g., annual refresher training
Skills	Pre-vocational, reinforced when required, e.g., after an adverse event	Pre-vocational, reinforce after experienced gained in the workplace
Attitudes	Pre-vocational	On entry to the workforce, reinforced regularly, e.g., annual refresher training
How?	**Awareness Training**	**Team Training**
Knowledge	Make available online/on demand	Make available online/on demand
Skills	Choose whether to 'make available' or impose	Conduct regular simulation training practising teamwork, and decision-making under pressure
Attitudes	Find champions to set an example	Institute review of case studies, regular discussion, mentoring, coaching

to consider sustainability: how do we maintain currency in skills that are not reinforced through practical application on a daily basis? To achieve this, we need to understand that learning is dynamic and must be reinforced with regular practice or refresher training. In addition, aligning other health care processes with the behavioural outcomes we desire from training, such as narrowing the gap between work-as-imagined and work-as-done throughout the system, will help.

15

Resilient Procedures: Oxymoron or Innovation?

Robert L. Wears and Garth S. Hunte

CONTENTS

Discussions about resilience often fetch up on the shoal of procedures[*] – prescriptive, feed-forward guidance. This is unfortunate because resilience is about a great deal more than procedures. In addition, the issue of procedures is often contested and acts as a 'black hole' sucking in all our attention and energy, while leaving little for other important areas of concern. This chapter discusses procedures in four parts: what baggage tends to accompany procedures; what is bad about procedures; what is good about them; and how procedures might be designed to support or even enhance resilience, instead of degrading it.

Contested Nature of Procedures

Procedures often come with excess baggage; issues that strictly speaking have little to do with the procedures themselves, but are social and political. Often, these 'extraneous' issues are dispositive, major contributors to the success or failure of attempts at proceduralization.

---

[*] In this chapter, we use the word 'procedures' to refer to a variety of types of feed-forward guidance: rules, regulations, guidelines, checklists, directives, doctrine, etc. Although distinctions among these exist, they are not relevant to this discussion.

Procedures are often the field upon which struggles for control play out. The popular 1969 film *Butch Cassidy and the Sundance Kid* opens with a contest between Butch and Harvey for the leadership of the Hole-in-the-Wall gang, but the struggle does not actually begin until the protagonists start to argue over the rules to be applied to their knife fight. Less fancifully, sociologists studying evidence-based medicine (a proceduralizing movement in health care) have characterized it as '... a move in the chess game of countervailing powers vying for dominance in the health care market ... [the] professional answer to the pressure from government agencies' and insurance companies' demands to render medicine more efficient and cost-effective' (Timmermans, 2010, p. 311). Thus what is presented as a merely technical exercise – just getting settled on a few procedures – is also a contest over control.

In addition, procedures may be produced for reasons other than providing guidance. They can be performances for external audiences, demonstrations that system leaders have adequate control of any hazards, and so used counter demands for external oversight. It is common in the aftermath of a celebrated accident for authorities to issue new rules to guarantee that 'this tragedy [will] never happen again' (Snook, 2000, p. 201). This commonly leads to idealized procedures that cannot possibly be followed. Clarke has called such procedures 'fantasy documents' (Clarke, 1999).

Procedures can also be useful foils for shifting responsibility for adverse events from the powerful (e.g., managers) to the less powerful (e.g., frontline workers) (Cook and Nemeth, 2010). 'Failure to follow procedure' is often noted as a cause of accidents, while 'poorly designed procedures' is almost never cited.

Finally, some of the appeal of procedures has been criticized as the technocratic wish, a reincarnation of scientism, '... modernity's rationalist dream that science can produce the knowledge required to emancipate us from scarcity, ignorance, and error' (Goldenberg, 2006, p. 2630), an objectivist presumption that the world and its technologies are ultimately knowable, and that epistemic ambiguity and uncertainty can be swept away (Kenny, 2014). This wish tends to imbue procedures with a power they do not have. For example, we see regularities in successful speech and from them deduce 'rules' of grammar – but these 'rules' are not the cause of successful speech, but rather the result of it (Scott, 2012).

Two Procedural Paradigms

Much of the controversy over the proper role of rules and procedures in promoting safe and effective work stems from two fundamentally different conceptual models of these forms of guidance (Dekker, 2005; Hale and

Borys, 2013b). The first model is rationalist, prescriptive and top-down; it is rooted in Taylorism. It envisions procedures as being devised by experts, in advance, away from the time and production pressures of the front lines. They are aimed at overcoming frailties of fallible operators, and tend to have a 'one and done' character, and seldom need to be modified. In this view, procedures precede work and guide it. This model tends to be dominant among managers and regulators, and is often explicitly re-articulated after an accident.

The second model sees procedures as socially constructed, locally situated and bottom-up; it is rooted in sociology and work ecology. It envisions procedures as emerging from work experience, and recognizes that they are essentially incomplete and require translation and adaptation to any specific situation. In this view, procedures follow work and are guided by it. This model tends to be more hidden from view; workers often hide work-as-done to avoid criticism or preserve resources; in addition, it may be hidden from the workers themselves in learned intuition.

Hale and Borys (2013a) have proposed a framework for bridging the gap between these disparate concepts of rules and procedures; we draw heavily on this framework as a model for crafting resilient procedures.

Problems with Procedures

Social and political issues aside, there are more intrinsic problems with procedures. Procedures can be seen as a specific case of standardization (Wears, 2015), and so they inherit all its associated advantages and disadvantages.

Procedures are necessarily decontextualized: abstractions of what generally is done, but not necessarily what specifically should be done in every situation. Complex organizations always run in mildly degraded states – someone is always absent, some device is inevitably out of service, some supplies are missing; workers normally must adapt procedures that presume ideal conditions to the actual conditions they face. If this context sensitivity is important to success, then slavish adherence to decontextualized procedures would be risky.

Attempts to compensate for this problem are ultimately self-defeating, for two reasons. First, there is a general problem with formalizations (of which procedures are a specific instance) in that they make some kinds of invisible work visible, but in so doing create a residue of implicit work that is not spelled out (Strauss et al., 1997). Second, this abstraction is actually part of their utility. A map is useful because of what it leaves out as much as by what it shows; a procedure covering everything would not be able to accomplish anything.

Indeed, a strong critique has developed arguing that some industries are over-proceduralized (Bieder and Bourier, 2013). A proliferation of procedures in ever greater detail becomes both internally contradictory (Shiffman, 1997) and impossible to grasp. You cannot improve safety and performance by thickening the rule-book (Bosk, 2006).

This problem points out an important precondition for procedures to be successful. Because they are decontextualized and abstracted, they depend heavily on the 'learned intuition' of the workers using them; the experiential knowledge of how things work and what is intended in context. Because all of us know how to open a jar, a procedure for making a peanut butter sandwich can pass over the multiple complex, coordinated steps involved in opening a jar. But this common grounding is seldom explicitly acknowledged or even recognized by procedure writers or users. (For example, the first step in the procedure of reaching forward to press an elevator button while standing has nothing to do with the arms; it is a tensioning of the gastrocnemius muscles of the legs to counteract the coming imbalance caused by moving an arm [roughly 8 pounds] forward of the body's centre of gravity).

Although it could theoretically be specifically included as an element of procedures (as noted below), many procedures in practice tend to deny the existence of trade-offs among goals, placing workers in a classic double bind based on the outcome of events, not on the adequacy of their judgement in negotiating trade-offs. Any trade-off conflicts that procedure writers are not able to resolve are pushed onto the front-line worker.

Of course in open systems, there will always be examples of situations in which following procedures led to disaster that deviation might have avoided. Vicente reports a simulation exercise in a nuclear power plant where the operators realized that the formal procedures for addressing a fault would eventually put the plant back into the original fault condition; so they deviated from the procedures in order to stabilize the plant safely. They were instructed that this was wrong, and they should instead adhere to procedures. When they did this in the next simulation, the plant was caught in an endless loop (i.e., the ostensibly corrective procedures returned it to its original fault condition); the operators were then cited for 'malicious compliance' (Vicente, 1999).

Benefits of Procedures

Procedures are generally viewed by their critics as intrinsically restraining, but there is no reason they need to be. The musical directives in Figure 15.1 illustrate this point. The musician is directed by the '*ff*' to play very loud, by the '*dim*' (for *diminuendo*) to gradually get softer, but also by the '*ad libitum*' to take liberties with the score – playing the notes exactly as written here would be to play them incorrectly, because here the composer wants performance variability.

Procedures have some intrinsic advantages. At their best, they can be viewed as distillations of experience, expertise and tacit, vernacular

FIGURE 15.1
Procedural directives in a musical score.

knowledge accumulated over many years by many workers in many situations. This can be particularly useful in seldom-encountered situations, particularly those situations in which the best action may be counter-intuitive.

Good procedures can serve as resources for situated actions. Even in off-base situations, procedures can provide 'a place to start', a way to make sense of an unfolding situation by taking actions and observing the responses. By automating, in a sense, some bits of work, they allow workers to direct scarce energy and attentional resources to its more important or more demanding aspects. In this sense they are particularly valuable as guides for novices.

When procedures are well adapted to context, well understood and generally observed, they can promote coordination at a distance and reduce the burden of articulation work by providing multiple actors an understanding of what their colleagues are likely to be doing. Similarly, by providing common ground, they promote efficiency in communication.

Schulman's study of the Diablo Canyon nuclear power plant provided an example of the effective use of a high degree of proceduralisation in promoting safe and effective operations (Schulman, 1993). Interestingly, the high degree of procedural formality was not accompanied by centralisation of authority, but rather by the reverse. By requiring a myriad of technical sub-units to have formalized procedures, and by giving each unit effective veto power over others' procedures, the plant increased the variety of viewpoints engaged in producing procedures, decreased structural secrecy and diminished the gap between work-as-imagined and work-as-done. By requiring frequent re-writing of procedures (often in one unit occasioned by another unit's procedural change) they were better able to keep pace with technological and environmental change.

Better Procedures as a Substrate for Resilience

Given this potential for procedures to be either good or bad, how might we craft better procedures that would support or even enhance, rather than degrade, resilience? This will require attention to two aspects of procedures: process and content.

Process

Process issues involve promoting conditions that provide the stage for, but do not necessarily ensure, better procedures. Increasing the variety of skills and viewpoints involved in procedure development, particularly if accompanied by flattening the organisation's hierarchy and shifting towards distributed, polycentric control enables an organisation to see more options and to identify more potential problems. This is particularly important when the procedures being developed are aimed at dealing with circumstances that are rarely, if ever, experienced (March et al., 1991).

Careful monitoring of the gap between work-as-imagined in the procedures and work-as-done at the 'coalface', coupled with a stance towards viewing an increasing gap as a sign that revision, not enforcement, is required, can help keep procedures practicable and abreast of changes in the work and the environment. This requires a commitment, not just lip service, to the Principle of Local Rationality: people only do things that make sense to them. Thus if people are deviating from the procedures, that is because they do not make sense to them, given their locally understood resources, constraints and goals. Careful monitoring of rule effectiveness in this way would allow organisations to properly enforce good and effective procedures, while redesigning or abandoning bad, unworkable or unnecessary procedures. Because procedures tend to accumulate but seldom go away (Townsend, 2013), regular reviews of effectiveness and necessity can help to combat this 'rule bloat'.

Interestingly, although many systems train their workers in procedures, they seldom train for adaptations of procedures. Discussions of the circumstances in which performance variability is necessary or desirable, and what good adaptations might look like would make procedures more flexible and more likely to be interpreted and applied safely and appropriately.

Finally procedure development should routinely employ methods to assess effectiveness prior to implementation. A variety of methods could be applied here: heuristic evaluation, formal assessment of completeness and consistency, table-top 'dry runs' or full-scale simulation. It is important that these assessments involve a requisite variety of viewpoints, positions and skills.

Content

On the content side, it would be useful to adopt the principle of equifinality: in complex systems, there are often many paths to the same goal. This is the antithesis of the Taylorist belief that there must always be 'one best way' to manage a work situation. Particularly under conditions of uncertainty, resource scarcity or time pressure, organisations cannot be certain that any single path will always be available to them, so they would do well to maintain multiple possible paths in a sort of 'portfolio strategy' as a hedge against uncertainty. A suite of 'pretty good' rules is likely to outperform a single 'best' rule in systems with ambiguous and uncertain inputs or working conditions.

Hale and Swuste have usefully advanced thinking about how to craft procedures wisely in their conception of goal, process and action rules (Hale and Swuste, 1998). Goal-oriented procedures identify the end state desired, but leave the selection of means unspecified. They provide useful guidance to resolving goal conflicts, particularly in intractable systems (Hollnagel, 2014b) where the object is not invariant performance, but rather the successful management of fluctuations. A simple example in health care comes from the principle when treating the two patients (mother and foetus) in a pregnancy, that 'the mother's life comes first'. This is not to say that mothers should never be exposed to risk to benefit the baby, but that when the two come directly into conflict, the best way to save the baby is to save the mother first. The specific actions to do so are left up to the actors in context; the procedure informs them about how to manage potential goal conflicts.

Process-oriented procedures do not specify actions, but rather how actions should be chosen, who should do the choosing and the factors they should consider (and conversely, which they should ignore). These procedures are useful in situations that still require flexibility in response, or as a means of coordination in polycentric control. For example, many systems have process rules for determining who has the authority to call in additional staff, and what circumstances should influence that decision.

The most detailed and specific procedures are action procedures, which list in detail which specific actions should be taken, by whom, in what order. Action procedures are typically what come to mind when we think of procedures; Hale and Swuste's contribution is significant in broadening our conception of what constitutes a procedure and thereby increasing procedures' utility. Clearly, action rules require both tractable systems (Hollnagel, 2014b) and stable environments to function properly.

Grote has noted that in high-risk systems, the most appropriate rules are not those that are most restrictive and therefore uncertainty reducing, but rather are flexible rules that promote adaptive action by providing more degrees of freedom to decision makers and thus increasing, rather than decreasing, their uncertainty (Grote, 2015).

There are some examples of flexible procedures in practice. One of the best is the German army's *Auftragssystem* from the Franco-Prussian War through World War II, which emphasized the primacy of the mission (goals) over orders (procedures) (Finkel, 2011). 'Far from being a horde of rigid and inflexible robots (… one of the most insidious of the Allied misconceptions about the Germans), the Wehrmacht … had a far greater ability to react or regain the initiative – especially in a moment of reverse – than was possessed particularly by the British Army of 1944' (Horne and Montgomery, 1994). This system provided explicitly that a subordinate commander should be able to pursue the tactical goals of his superiors with or without orders, based on his knowledge of the local context (Reason, 1997). This of course presupposes a considerable degree of initiative and tactical understanding on the part of these junior leaders, which points out that supportive procedures are only one component of a resilient system; another is developing the capability for responsive adaptation in front-line leaders. The success of these flexible procedures showed on the battlefield; whether attacking or defending, the German army inflicted substantially more casualities than it suffered in these conflicts.

Conclusion

A good deal of the literature about resilience has concerned itself with phenomenology – detailed, thick descriptions of specific manifestations of resilient or brittle performance in demanding circumstances. However, this focus has been limiting because it fails to address the question of where those resilient performances come from; what substrate supports them. There must be something about a system that provides the foundation for resilience when called upon (Fairbanks et al., 2014). Much of a system's repertoire of responses is hidden in the form of latent behavioural potential, out of awareness and taken for granted until some triggering event calls attention to it (Christianson et al., 2009). If we want to understand how to engineer resilience into our systems, we must find ways of identifying, conserving and developing that potential. One way to do that is to learn how to design effective, flexible procedures that support resilience rather than interfere with it. This may require organizations to embrace uncertainty rather than to pursue the will-of-the-wisp of rationalist certainty.

16

Conclusion: Pathways Towards Reconciling WAI and WAD

Jeffrey Braithwaite, Robert L. Wears and Erik Hollnagel

Throughout these pages, the writers have first sought to appreciate the behavioural nuances of those who try to influence clinical work from a distance, and people who carry out everyday clinical activities in practice. They have also sought, second, to examine how these two worlds interface. And third, they have sought to redress a challenge that faces not just those in health care, but everywhere else: to what extent can the blunt end, representing work-as-imagined (WAI), and the sharp end, representing work-as-done (WAD), be reconciled?

Several things will strike readers as they absorb these contributions. One is the way stakeholders, whether sharp- or blunt-end operators, variously navigate their environment and juggle their tasks and responsibilities as they traverse their bit of the world, trying to make headway. Another is how, to make things work better, trust and reciprocity are needed not just *within* blunt-end or sharp-end groups, but *across* the WAI–WAD differences.

Yet another point worth underlining is that it is simply not feasible, in theory and in practice, that clinicians doing the work of health care could adhere to all the prescriptive rules, policies and procedures that are specified for them, especially in an era of bureaucratic impositions and rules bloat. Equally, it is not feasible that policymakers or managers could alter all their rules, policies and procedures such that they corresponded to WAD. So whatever we do in the future, if we are to have safer systems of care, we need to harness all the brains and expertise at our disposal. WAD and WAI exponents will need to become better at appreciating each other's expertise, listening to each other's point of view and making trade-offs, negotiating, power-sharing, collaborating and partnering. We would do well to remember the old proverb: *none of us is as smart as all of us.*

So, what have we learned from these contributions to enable such accommodations? Table 16.1 draws out some salient points from the wide-ranging solutions to the task we set our authors and ourselves of trying to enjoin the otherwise circumscribed worlds of WAI and WAD.

TABLE 16.1

Synthesis of Authors' Key Contributions and Lessons Learned

Authors, Chapter	Selected Key Lessons	Country	Empirical Stance	Theoretical Approach
Braithwaite, Wears, Hollnagel Preface	Previous volumes in the trilogy of this work focused on resilience and everyday clinical work. Now, we are moving to a consideration of how sharp-end (WAD) and blunt-end (WAI) interests can be conciliated with a view to providing safer, more appropriate care.	Australia, United States, Denmark	Presentation of prior case study work	Resilience theory
Hollnagel Prologue: Why Do Our Expectations of How Work Should Be Done Never Correspond Exactly to How Work *Is Done?*	Expectations or descriptions of work can never match the complexity of work actually done. Suggestions for tackling this problem include learning more about work-in-practice, encouraging diversity of perspectives and reflecting mindfully on the gaps between WAI and WAD.	Denmark	Reliance on wide-ranging literature	Resilience theory
Part I: Problems and Issues				
Saurin, Rosso, Colligan Chapter 1: Towards a Resilient and Lean Health Care	Although resilience engineering and Lean production represent two different approaches to health systems management, their approaches can be used synergistically. Both can illuminate aspects of WAI and WAD.	Brazil, United States	Methodological comparison of FRAM and VSM	Theories of implementation and application
Sheps, Cardiff Chapter 2: The Jack Spratt Problem: The Potential Downside of Lean Application in Health Care – A Threat to Safety-II	The impact of implementing a Lean approach in health care requires careful consideration. Health care is inherently complex and a Lean approach often does not allow for flexibility in WAD; creating leaner services may have unintended consequences for patient safety.	Canada	Comparison using three case examples	Continuous improvement theory

Canfield Chapter 3: Recovery to Resilience: A Patient Perspective	After her husband died as a result of an adverse event, Canfield advocates for more trust in relationships between health professionals. Including the patient and their family as part of the team and as active contributors to patient safety is a way forward.	Canada	Personal insights, experience and expertise	The patient-as-knowledge- and change-broker
Nyssen, Bérastégui Chapter 4: Is System Resilience Maintained at the Expense of Individual Resilience?	To strengthen resilience, self-reflection, self-regulation and self-determination must be prominent. This contrasts with an approach that unduly emphasises ideal work, prescribed work or mandated work.	Belgium	Studies of doctors and police officers	Theories of individual and collective resilience
Chuang, Hollnagel Chapter 5: Challenges in Implementing Resilient Health Care	Stress–strain plots, resilience analysis grids, the functional resonance analysis method and other approaches used in resilience engineering can be and are being employed in health systems management to reconcile WAI and WAD. This can help facilitate health care system resilience. Having a holistic, practical, conceptual model of resilient health care is important.	Taiwan, Denmark	Reduction of bloodstream infections in central lines in intensive care units	Implementation science model of improvement

Part II: Applications

Nakajima, Masuda, Nakajima Chapter 6: Exploring Ways to Capture and Facilitate Work-as-Done That Interacts with Health Information Technology	Using a FRAM model, the authors identify control signals from technologies to clinicians as being a key target for systemic health improvement. Compliance-response models are insufficient, and a poor match to how clinical work is done in practice.	Japan	Surveys of the use of IT to identify patients and manage blood transfusions and medication administration	FRAM analysis

(Continued)

TABLE 16.1 (*Continued*)

Synthesis of Authors' Key Contributions and Lessons Learned

Authors, Chapter	Selected Key Lessons	Country	Empirical Stance	Theoretical Approach
Johnson, Lane Chapter 7: Resilience Work-as-Done in Everyday Clinical Work	The authors propose the Ten C model, a multidimensional, interconnected model for understanding, describing, teaching and implementing resilience in WAD: cohesion, capture, cognition, communication, culture, clear ownership, constraints, challenge, competence and compliance. Cohesion, or mutual respect, is key for delivering safe and efficient health care.	Australia	Model arising from stakeholder interactions	Model-application theory
Braithwaite, Clay-Williams, Hunte, Wears Chapter 8: Understanding Resilient Clinical Practices in Emergency Department Ecosystems	An exploration of emergency care case studies shows that aligning WAI and WAD requires high levels of trust and reciprocity. It also needs explicit knowledge sharing and an understanding of other stakeholders' perspectives.	Australia, Canada, United States	Comparisons of five case studies in emergency departments	Theories of organisational complexity
Sujan, Pozzi, Valbonesi Chapter 9: Reporting and Learning: From Extraordinary to Ordinary	Learning from ordinary, everyday clinical work, rather than learning exclusively from the extraordinary (e.g., adverse events) can build an understanding of WAD. It can help expose the decisions and modifications clinicians make in order to deliver safe care.	United Kingdom, Italy	Case study of the proactive risk monitoring in health care (PRIMO) approach	Learning theory
Cook and Ekstedt Chapter 10: Reflections on Resilience: Repertoires and System Features	This is a reflective contribution on resilience and how it occurs naturally in systems. Resilience is present in systems regardless of attempts to engineer them. Having the ability to engineer resilience means we also have the ability to erode it.	United States, Sweden	Systems analysis	Resilience engineering theory

Hunte, Wears Chapter 11: Power and Resilience in Practice: Fitting a 'Square Peg in a Round Hole' in Everyday Clinical Work	A consideration of WAI and WAD from a perspective of power. Argues for a generative partnership of top-down and bottom-up approaches, rather than one or the other.	Canada, United States	Emergency department case exemplar	Theories of power and enactment

Part III: Methods and Solutions

Anderson, Ross, Jaye Chapter 12: Modelling Resilience and Researching the Gap between Work-as-Imagined and Work-as-Done	Presents a working model of organisational resilience to identify potential targets for intervention to improve patient safety. Core concepts: WAI-demand and capacity; WAD-successes and failures. An understanding of the types and causes of pressures faced in clinical work, how they promote adaptive behaviour and how they interrelate is required.	United Kingdom	Theoretical model applied to systems understanding	Resilience engineering theory
Patterson, Deutsch, Jacobson Chapter 13: Simulation: Closing the Gap between Work-as-Imagined and Work-as-Done	Simulation can be used to understand and improve health care resilience. The benefits of simulation include that it connects WAI and WAD, moves us closer to WAD, and identifies the boundaries of performance.	United States	Functional examples of simulation in action	Simulation theory in the context of resilience engineering theory
Clay-Williams, Braithwaite Chapter 14: Realigning Work-as-Imagined and Work-as-Done: Can Training Help?	Awareness training and team training of leaders, clinicians, patients and their carers at various time points in their career through online training, case reviews, simulation training and mentoring may create a more resilient health care system and realign WAI with WAD.	Australia	Application of theoretical principles to health care settings	Teaching and learning theories

(Continued)

TABLE 16.1 (*Continued*)

Synthesis of Authors' Key Contributions and Lessons Learned

Authors, Chapter	Selected Key Lessons	Country	Empirical Stance	Theoretical Approach
Wears, Hunte Chapter 15: Resilient Procedures: Oxymoron or Innovation?	Developing effective and flexible procedures (including policies, checklists, guidelines) that support flexibility and adjustment rather than constrain it can help us strengthen resilience in health systems.	United States, Canada	Systems analysis of the application of theories	Theories of procedures and adoption
Braithwaite, Wears, Hollnagel Chapter 16: Conclusion: Pathways towards Reconciling WAI and WAD	Both WAI and WAD are essential to provide effective health care. Working on ways over time to reconcile the two seems generally to be a good use of time and expertise.	Australia, United States, Denmark	Critical synthesis of prior material	Theoretical pluralities

Armed with Table 16.1 as a summary, we can now see how the jigsaw puzzle pieces of the book fit together as a blueprint for the future. Across the 18 contributions (16 chapters, a Preface and a Prologue), we have 11 countries represented by 30 authors embracing a wide range of empirical positions and theoretical perspectives. Different countries, different settings, different thought leaders, different empirical traditions and different theories of how the world works are being mobilised in pursuit of this enterprise: to reconcile WAI and WAD.

The table represents the final take-home message with which we would leave readers. There are many pathways to improved health care, to build more resilient systems and to reconcile the two worldviews. In making their respective contributions to performance, WAI and WAD can indeed help make things safer for patients.

So this collection of chapters points to the fact that it is always unwise in a complex world to think there is one way, a best method, or an optimal solution. There is not. But there are smart ideas, new stances, fresh perspectives, additional studies and novel recommendations and suggestions. Although the tensions between WAI (which helps stabilise systems of work and combat sub-optimisation) and WAD (which helps work systems respond to local contingencies) are omnipresent and never definitively resolvable, we aim to have shown that they can be actively, creatively and respectfully managed in context (Greenhalgh et al., 2009).

The three books have explored new ways to think about clinical care and make it safe, to support resilience in health care settings, to strengthen care for patients and to make progress in figuring out how to reconcile WAI and WAD. At the end of the day we and our authors will hopefully have teased out, deconstructed and illuminated these worlds, and presented new, old and re-assembled ideas as well as tendered solutions to problems facing blunt- and sharp-end participants. In short, we hope we have made a badly needed contribution to future health care.

References

Alonso, A. et al. (2006). Reducing medical error in the military health system: How can team training help? *Human Resource Management Review,* 16(3), 396–415.

Amalberti, R. and Deblon, F. (1992). Cognitive modelling of fighter aircraft process control: A step towards an intelligent on-board assistance system. *International Journal of Man-Machine Studies,* 36, 639–671.

American Association of Neurological Surgeons. (2014). *Tethered spinal cord syndrome.* Online at http://www.aans.org/Patient Information/Conditions and Treatments/Tethered Spinal Cord Syndrome.aspx, accessed 5 March 2015.

American Hospital Association. (2003). *The Patient Care Partnership.* Online at http://www.aha.org/advocacy-issues/communicatingpts/pt-care-partnership.shtml, accessed 24 December 2014.

Anaut, M. (2003). *La Résilience: Surmonter les Traumatismes.* Nathan Université, Saint Germain–du-Puy, France: Armand Colin.

Anders, S. et al. (2011). Blood product positive patient identification: Comparative simulation-based usability test of two commercial products. *Transfusion,* 5(11), 2311–2318.

Anderson, J. E. et al. (2013). Can incident reporting improve safety? Healthcare practitioners' views of the effectiveness of incident reporting. *International Journal for Quality in Health Care,* 25, 141–150.

Anonymous. (2014). Schumpeter: The holes in holacracy. *The Economist.* 14 July. Online at http://www.economist.com/node/21606267/print, accessed 15 July 2015.

Antonsen, S. (2009). Safety culture and the issue of power. *Safety Science,* 47(2), 183–191.

Araz, O. M. et al. (2012). A new method of exercising pandemic preparedness through an interactive simulation and visualization. *Journal of Medical Systems,* 36(3), 1475–1483.

Arendt, H. (1998). *The Human Condition* (2nd edition). Chicago, IL: University of Chicago Press.

Argyris, C. and Schön, D. A. (1996). *Organizational Learning II: Theory, Method and Practice.* Reading, MA: Addison-Wesley.

Ashby, W. R. (1956). *An Introduction to Cybernetics.* London, UK: Methuen.

Asplin, B. R. et al. (2003). A conceptual model of Emergency Department overcrowding. *Annals of Emergency Medicine,* 42(2), 173–180.

Association of American Medical Colleges. *Readiness for Reform. Virginia Mason Medical Center: Applying LEAN Methodology to Lead Quality and Transform Healthcare.* Online at https://www.aamc.org/download/278946/data/virginia-masoncasestudy.pdf, accessed 31 December 2014.

Auerbach, M., Kessler, D. and Patterson, M. (2015). The use of in-situ simulation to detect latent safety threats in pediatrics: An international descriptive cross-sectional survey. *BMJ Simulation and Technology Enhanced Learning,* 1(3), 77–82, first published online: 3 November.

Australian Commission on Safety and Quality in Health Care. (2008). *Australian Charter for Healthcare Rights.* Online at http://www.safetyandquality.gov.au/national-priorities/charter-of-healthcare-rights/, accessed 24 December 2014.

Baker, D. P. et al. (2003). *Medical Teamwork and Patient Safety: The Evidence-Based Relation*. Washington, DC: American Institute for Research.

Baker, D. P. et al. (2005). *Medical Team Training Programs in Health Care. Advances in Patient Safety: From Research to Implementation*. AHRQ Publication Nos. 050021 (1–4). Rockville, MD: Agency for Healthcare Research and Quality.

Baker, D. P. et al. (2010). Assessing teamwork attitudes in healthcare: Development of the Team STEPPS teamwork attitudes questionnaire. *Quality and Safety in Health Care*, 19(6), e49.

Bakhtin, M. M. (2010). *The Dialogic Imagination: Four Essays*. Austin, TX: University of Texas Press.

Banja, J. D. (2004). *Medical Errors and Medical Narcissism*. Sudbury, MA: Jones and Bartlett.

Barach, P. and Johnson, J. K. (2006). Understanding the complexity of redesigning care around the clinical microsystem. *Quality and Safety in Health Care*, 15(suppl. 1), i10–i16.

Barach, P. and Small, S. D. (2000). Reporting and preventing medical mishaps: Lessons from non-medical near miss reporting systems. *BMJ*, 320, 759–763.

Barber, N. (2002). Should we consider non-compliance a medical error? *Quality and Safety in Health Care*, 11(1), 81–84.

Benn, J. et al. (2009). Feedback from incident reporting: Information and action to improve patient safety. *Quality and Safety in Health Care*, 18, 11–21.

Bérastégui, P. (2014). *Emergency Services: Analyse de L'activité et Extraction de Compétences Transversales*. Unpublished Master Thesis in Psychological Sciences, Université de Liège, Belgium.

Bergström, J. et al. (2011). Training organisational resilience in escalating situations. In E. Hollnagel et al. (eds.), *Resilience Engineering in Practice: A Guidebook*. Farnham, Surrey, UK: Ashgate (pp. 45–56).

Berwick, D. M. (2009). What 'patient-centered' should mean: Confessions of an extremist. *Health Affairs*, 28(4), w555–w565.

Besnard, D. and Hollnagel, E. (2014). I want to believe: Some myths about the management of industrial safety. *Cognition, Technology and Work*, 16(1), 13–23.

Bieder, C. and Bourier, M. (eds.), (2013). *Trapping Safety into Rules: How Desirable or Avoidable Is Proceduralization*. Farnham, Surrey, UK: Ashgate.

Black, J. R. (2008). *The Toyota Way to Healthcare Excellence: Increase Efficiency and Improve Quality with Lean*. Chicago, IL: Health Administration Press.

Bosk, C. L. (2006). All things twice, first tragedy then farce: Lessons from a transplant error. In K. Wailoo, J. Livingston and P. Guarnaccia (eds.), *A Death Retold*. Chapel Hill, NC: University of North Carolina Press (pp. 97–116).

Braithwaite, J. (2010). Between-group behaviour in health care: Gaps, edges, boundaries, disconnections, weak ties, spaces and holes. A systematic review. *BMC Health Services Research*, 10, 330.

Braithwaite, J. et al. (2010). Cultural and associated enablers of, and barriers to, adverse incident reporting. *Quality and Safety in Health Care*, 19, 229–233.

Braithwaite, J. et al. (2012). A four-year, systems-wide intervention promoting interprofessional collaboration. *BMC Health Services Research*, 12, 99–106.

Braithwaite, J. et al. (2013a). Health care as a complex adaptive system. In E. Hollnagel, J. Braithwaite and R. L. Wears (eds.), *Resilient Health Care*. Farnham, Surrey, UK: Ashgate (pp. 57–76).

Braithwaite, J. et al. (2013b). Continuing differences between health professions' attitudes: The saga of accomplishing systems-wide interprofessionalism. *International Journal for Quality in Health Care*, 25, 8–15.

Braithwaite, J., Hyde, P. and Pope, C. (2010). *Culture and Climate in Health Care Organizations*. Farnham, Surrey, UK: Ashgate.

Brandão de Souza, L. and Pidd, M. (2011). Exploring the barriers to lean health care implementation. *Public Money and Management*, 31, 59–66.

Buchanan, D. A. (1997). The limitations and opportunities of business process re-engineering in a politicized organizational climate. *Human Relations*, 50(1), 51–72.

Buljac-Samardzic, M. et al. (2009). Interventions to improve team effectiveness: A systematic review. *Health Policy*, 94(3), 183–195.

Bullard, M. J. et al. (2012). The role of a rapid assessment zone/pod on reducing overcrowding in Emergency Departments: A systematic review. *Emergency Medicine Journal*, 29(5), 372–378.

Cannon-Bowers, J. A. and Salas, E. E. (1998). *Making Decisions Under Stress: Implications for Individual and Team Training*. Washington, DC: American Psychological Association.

Carrico, R. and Ramirez, J. (2007). A process for analysis of sentinel events due to health care-associated infection. *American Journal of Infection Control*, 35, 501–507.

Carruthers, I. and Phillip, P. (2006). *Safety First: A Report for Patients, Clinicians and Healthcare Managers*. London, UK: National Patient Safety Agency.

Carthey, J. et al. (2011). Breaking the rules: Understanding non-compliance with policies and guidelines. *BMJ*, 343, d5283.

Carver, C. (1997). You want to measure coping but your protocol's too long: Consider the brief cope. *International Journal of Behavioural Medicine*, 4(1), 92–100.

Chandler, D. (2014). *Resilience: The Governance of Complexity*. New York, NY: Routledge.

Christianson, M. K. et al. (2009). Learning through rare events: Significant interruptions at the Baltimore and Ohio Railroad Museum. *Organization Science*, 20(5), 846–860.

Chuang, S. & Wears, R. L. (2015). Strategies to get resilience into everyday clinical work. In R. L. Wears, E. Hollnagel and J. Braithwaite (eds.), *Resilient Health Care: The resilience of everyday clinical work. vol. 2*. Farnham, Surry, UK: Ashgate (pp. 225–235).

Cilliers, P. (1998). *Complexity and Postmodernism: Understanding Complex Systems*. New York, NY: Routledge.

Clarke, L. (1999). *Mission Improbable: Using Fantasy Documents to Tame Disaster*. Chicago, IL: University of Chicago Press.

Clay-Williams, R. (2010). *Multidisciplinary Crew Resource Management (CRM) in Health Care: Attitude and Behaviour Change Associated with Classroom and Simulation-Based Training*. Unpublished Doctor of Philosophy Thesis, University of New South Wales, Sydney, Australia.

Clay-Williams, R. (2013). Re-structuring and the resilient organisation: Implications for health care. In E. Hollnagel, J. Braithwaite and R. L. Wears (eds.), *Resilient Health Care*. Farnham, Surrey, UK: Ashgate (pp. 123–133).

Clay-Williams, R. and Braithwaite, J. (2009). Determination of health-care teamwork training competencies: A Delphi study. *International Journal for Quality in Health Care*, 21(6), 433–440.

Clay-Williams, R. and Braithwaite, J. (2013). Safety-II thinking in action: 'Just in time' information to support everyday activities. In E. Hollnagel, J. Braithwaite and R. L. Wears (eds.), *Resilient Health Care*. Farnham, Surrey, UK: Ashgate (pp. 205–213).

Clay-Williams, R. and Braithwaite, J. (2014). Reframing implementation as an organisational behaviour problem: Inside a teamwork improvement intervention. *Journal of Health Organization and Management*, 29(6), 670–683.

Clay-Williams, R. et al. (2013). Classroom and simulation team training: A randomized controlled trial. *International Journal for Quality in Health Care*, 25, 314–321.

Clay-Williams, R. et al. (2014). On a wing and a prayer: An assessment of modularized Crew Resource Management training for health care professionals. *Journal of Continuing Education in the Health Professions*, 34, 56–67.

Clegg, S. (1975). *Power, Rule and Domination: A Critical and Empirical Understanding of Power in Sociological Theory and Organizational Life*. London, UK: Routledge & Kegan Paul.

Clegg, S. (1989). *Frameworks of Power*. London, UK: Sage.

Clot, Y. (1999). *Le Ravail à Cœur. Pour en Finir avec les Risques Psychosociaux*. Paris, France: La Découverte.

Coats, T. J. and Michalis, S. (2001). Mathematical modelling of patient flow through an accident and Emergency Department. *Emergency Medicine Journal*, 18, 190–192.

Cook, R. I. (1997). Observations on RISKS and risks. *Communications of the ACM*, 40(3), 122.

Cook, R. I. (1998). *How Complex Systems Fail*. Cognitive Technologies Laboratory. Online at http://web.mit.edu/2.75/resources/random/How%20Complex%20Systems%20Fail.pdf, accessed 31 December 2014.

Cook, R. I. (1999). Two years before the mast: Learning to learn about patient safety. In *Proceedings of Enhancing Patient Safety and Reducing Error in Health Care Conference*, National Patient Safety Foundation, Chicago, IL, 8–10 December.

Cook, R. I. (2013a). Resilience, the second story, and progress on patient safety. In E. Hollnagel, J. Braithwaite and R. L. Wears (eds.), *Resilient Health Care*. Farnham, Surrey, UK: Ashgate (pp. 19–26).

Cook, R. I. (2013b). Resilience in complex adaptive systems: Operating at the edge of failure. Presented at Velocity 2013. Online at https://www.youtube.com/watch?v=PGLYEDpNu60, accessed 19 March 2015.

Cook, R. I. and Nemeth, C. (2006). Taking things in stride: Cognitive features of two resilient performances. In E. Hollnagel, D. D. Woods and N. G. Leveson (eds.), *Resilience Engineering: Concepts and Precepts*. Aldershot, UK: Ashgate (pp. 205–221).

Cook, R. I. and Nemeth, C. (2010). "Those found responsible have been sacked": Some observations on the usefulness of error. *Cognition, Technology and Work*, 12(1), 87–93.

Cook, R. and O'Connor, M. (2005). Thinking About Accidents and Systems. In K. Thompson and H. Manasse (eds.), *Improving Medication Safety*. Washington, DC: ASHP.

Cook, R. I., Woods, D. D. and Miller, C. (1998). *A Tale of Two Stories: Contrasting Views of Patient Safety*. Chicago, IL: National Patient Safety Foundation.

Cooper, J. B. et al. (2011). Design and evaluation of simulation scenarios for a program introducing patient safety, teamwork, safety leadership, and simulation to healthcare leaders and managers. *Simulation in Healthcare*, 6, 231–238.

Cortwright, E. M. (1975). Apollo: Expeditions to the Moon. *National Aeronautics and Space Administration (NASA), Scientific and Technical Information Office, SP-350.* Washington, DC: NASA.

Creswick, N. and Westbrook, J. (2006). Examining the socialising and problem-solving networks of clinicians on a hospital ward. In *Conference Proceedings of Social Science Methodology Conference of the Australian Consortium for Social and Political Research (ACSPR)*, Sydney, Australia, 10–13 December.

Creswick, N., Westbrook, J. and Braithwaite, J. (2009). Understanding communication networks in the Emergency Department. *BMC Health Services Research, 9,* 247.

Crosskerry, P. (2002). Achieving quality in clinical decision making: Cognitive strategies and detection of bias. *Academic Emergency Medicine,* 9(11), 1184–1204.

Cyrulnik, B. (1999). *Un Merveilleux Malheur.* Paris, France: Odile Jacob.

Damiani, C. and Pereira-Fradin, M. (2006). *Traumaq: Questionnaire D'évaluation du Traumatisme Psychique.* Paris, France: ECPA.

Dan, H. et al. (2012). The ward rounds with in situ simulation to observe patient identification procedures in the administration of blood components using health information technology [in Japanese]. *Japanese Journal of Quality and Safety in Healthcare,* 17(suppl.), 235.

de Souza, L. B. (2009). Trends and approaches in Lean healthcare. *Leadership in Health Services,* 22(2), 121–139.

Debono, D. and Braithwaite, J. (2015). Workarounds in nursing practice in acute care: A case of a health care arms race? In R. L. Wears, E. Hollnagel and J. Braithwaite (eds.), *Resilient Health Care, Volume 2: The Resilience of Everyday Clinical Work.* Farnham, Surrey, UK: Ashgate (pp. 23–37).

Debono, D. S. et al. (2013). Nurses' workarounds in acute healthcare settings: A scoping review. *BMC Health Services Research,* 13, 175.

Dekker, S. W. A. (2003). Failure to adapt or adaptations that fail: Contrasting models on procedures and safety. *Applied Ergonomics,* 34(3), 233–238.

Dekker, S. W. A. (2005). *Ten Questions about Human Error: A New View of Human Factors and System Safety.* Mahwah, NJ: Lawrence Erlbaum.

Dekker, S. W. A. (2006). Resilience engineering: Chronicling the emergence of confused consensus. In E. Hollnagel, D. D. Woods and N. G. Leveson (eds.), *Resilience Engineering: Concepts and Precepts.* Farnham, Surrey, UK: Ashgate (pp. 77–92).

Dekker, S. W. A. (2011). *Drift into Failure: From Hunting Broken Components to Understanding Complex Systems.* Farnham, Surrey, UK: Ashgate.

Dekker, S. W. A. et al. (2013). Complicated, complex, and compliant: Best practice in obstetrics. *Cognition, Technology and Work,* 15(2), 189–195.

Dekker, S. W. A., Cilliers, P. and Hofmeyr, J. H. (2011). The complexity of failure: Implications of complexity theory for safety investigations. *Safety Science,* 49, 939–945.

DelliFraine, J. L., Langabeer, J. R. and Nembhard, I. (2010). Assessing the evidence of Six Sigma and LEAN in the healthcare industry. *Quality Management in Health Care,* 19(3), 211–225.

Demaria, S. Jr. et al. (2010). Adding emotional stressors to training in simulated cardiopulmonary arrest enhances participant performance. *Medical Education,* 44(10), 1006–1015.

Department of Health. (2000). *An Organisation with a Memory.* London, UK: Stationary Office.

Dixon-Woods, M. et al. (2014). *Safer Clinical Systems: Evaluation Findings*. London, UK: Health Foundation.

Donyai, P. et al. (2008). The effects of electronic prescribing on the quality of prescribing. *British Journal of Clinical Pharmacology*, 65, 230–237.

Doyle, M. J. and Marsh, L. (2013). Stigmercy 3.0: From ants to economics. *Cognitive Systems Research*, 21(1), 1–6.

Driskell, J. E. et al. (2008). Stress exposure training: An event-based approach. In P. Hancock and J. Szalma (eds.), *Performance Under Stress*. Aldershot, UK: Ashgate (p. 271).

Driskell, J. E., Johnston, J. H. and Salas, E. (2001). Does stress training generalize to novel settings? *Human Factors*, 43(1), 99–110.

Duit, A. et al. (2010). Governance, complexity, and resilience. *Global Environmental Change*, 20(3), 363–368.

Dujarier, M.-A. (2006). *L'idéal au Travail*. Paris, France: Presses Universitaires de France.

Duncan, R. B. (1974). Modifications in decision structure in adapting to the environment: Some implications for organizational learning. *Decision Science*, 5, 705–725.

Edwards, A. and Elwyn. G. (2009). *Shared Decision-Making in Health Care: Achieving Evidence-Based Patient Choice*. Oxford, UK: Oxford University Press.

Eksted, M. and Cook, R. I. (2015). The Stockholm blizzard of 2012. In R. L. Wears, E. Hollnagel and J. Braithwaite (eds.), *Resilient Health Care, Volume 2: The Resilience of Everyday Clinical Work*. Farnham, Surrey, UK: Ashgate (pp. 59–72).

Epstein, S. (2008). Unexampled events, resilience and PRA. In E. Hollnagel, C. P. Nemeth and S. W. A. Dekker (eds.), *Resilience Engineering Perspectives, Volume 1: Remaining Sensitive to the Possibility of Failure*. Farnham, Surrey, UK: Ashgate (pp. 49–62).

Esimai, G. (2005). Lean Six Sigma reduces medication errors. *Quality Progress*, 38(4), 51–57.

European Commission. (2002). *European Charter of Patients' Rights*. Online at http://ec.europa.eu/health/ph_overview/co_operation/mobility/docs/health_services_co108_en.pdf, accessed 24 December 2014.

Fairbanks, R. J. et al. (2013). Separating resilience from success. In E. Hollnagel, J. Braithwaite and R. L. Wears (eds.), *Resilient Health Care*. Farnham, Surrey, UK: Ashgate (pp. 1162–1164).

Fairbanks, R. J. et al. (2014). Resilience and resilience engineering in health care. *Joint Commission Journal on Quality and Patient Safety*, 40(8), 376–383.

Farjoun, M. (2010). Beyond dualism: Stability and change as a duality. *Academy of Management Review*, 35(2), 202–225.

Finkel, M. (2011). *On Flexibility: Recovery from Technological and Doctrinal Surprise on the Battlefield*. Stanford, CA: Stanford University Press.

Fiol, C. M. and Lyles, M. A. (1985). Organizational learning. *Academy of Management Review*, 10, 803–813.

Fleming, P. and Spicer, A. (2014). Power in management and organization science. *Academy of Management Annals*, 8(1), 237–298.

Flin, R., O'Connor, P. and Crichton, M. (2008). *Safety at the Sharp End: A Guide to Non-technical Skills*. Aldershot, UK: Ashgate.

Flyvbjerg, B. (2001). *Making Social Science Matter: Why Social Enquiry Fails and How It Can Succeed Again*. Cambridge, UK: Cambridge University Press.

Folke, C. et al. (2005). Adaptive governance of social-ecological systems. *Annual Review of Environment and Resources*, 30, 441–473.

Forero, R. and Hillman K. M. (2008). *Access Block and Overcrowding: A Literature Review.* Report prepared for the Australian College of Emergency Medicine (ACEM). Simpson Centre for Health Services Research. University of New South Wales, Sydney, Australia.

Foucault, M. (1975). *Discipline and Punish*. New York, NY: Random House.

Frain, J. et al. (2004). *Integrating Sentinel Event Analysis into Your Infection Control Practice.* White paper from the Association for Professionals in Infection Control and Epidemiology (APIC). Online at http://www.apic.org/Resource_/TinyMceFileManager/Position_Statements/Sentinel-Event.pdf, accessed 23 January 2015.

Franc-Law, J. M., Bullard, M. and Della Corte, F. (2008). Simulation of a hospital disaster plan: A virtual, live exercise. *Prehospital and Disaster Medicine*, 23(4), 346–353.

Frank, A. W. (2004). *The Renewal of Generosity: Illness, Medicine, and How to Live.* Chicago, IL: University of Chicago Press.

Fraser, K. et al. (2012). Emotion, cognitive load and learning outcomes during simulation training. *Medical Education*, 46(11), 1055–1062.

Freeman, M., Miller, C. and Ross, N. (2000). The impact of individual philosophies of teamwork on multi-professional practice and the implications for education. *Journal of Interprofessional Care*, 14(3), 237–247.

Furniss, D. et al. (2011). A resilience markers framework for small teams. *Reliability Engineering and System Safety*, 96(1), 2–10.

Gardner, A. K. et al. (2013). In situ simulation to assess workplace attitudes and effectiveness in a new facility. *Simulation in Healthcare*, 8(6), 351–358.

Gauthereau, V. and Hollnagel, E. (2005). Planning, control, and adaptation: A case study. *European Management Journal*, 23(1), 118–131.

Geidl, L. et al. (2011). Intuitive use and usability of ventricular assist device peripheral components in simulated emergency conditions. *Artificial Organs*, 35(8), 773–780.

Geis, G. L. et al. (2011). Simulation to assess the safety of new healthcare teams and new facilities. *Simulation in Healthcare*, 6(3), 125–133.

Gerberding, J. L. (2002). Hospital-onset infections: A patient safety issue. *Annals of Internal Medicine*, 137, 665–670.

Ghanbarzadeh, R. et al. (2014). A decade of research on the use of three-dimensional virtual worlds in health care: A systematic literature review. *Journal of Medical Internet Research*, 16(2), e47.

Giddens, A. (1984). *The Constitution of Society*. Cambridge, UK: Polity Press.

Gittell, J. H. (2009). *High Performance Healthcare: Using the Power of Relationships to Achieve Quality, Efficiency and Resilience.* New York, NY: McGraw-Hill.

Goldenberg, M. J. (2006). On evidence and evidence-based medicine: Lessons from the philosophy of science. *Social Science and Medicine*, 62(11), 2621–2632.

Goldratt, E. (1984). *The Goal: A Process of Ongoing Improvement.* Great Barrington, MA: North River Press.

Goldstein, I. L. and Ford, J. K. (2002). *Training in Organizations: Needs Assessment, Development, and Evaluation.* Belmont, CA: Wadsworth.

Government of Saskatchewan. (2014). *Innovations in Health Care: Lean, Saskatchewan Health Care Management System.* Online at http://www.health.gov.sk.ca/lean, accessed 31 December 2014.

Graafland, M., Schraagen, J. and Schijven, M. (2012). Systematic review of serious games for medical education and surgical skills training. *British Journal of Surgery*, 99, 1322–1330.

Grassé, P. P. (1959). La reconstruction du nid et les coordinations interindividuelles chez bellicositermes natalensis et cubitermes sp. la thorie de la stigmergie: Essai dinterprtation du comportement des termites constructeurs. *Insectes Sociaux*, 6(1), 41–83.

Greenfield, D. et al. (2009). Distributed leadership to mobilise capacity for accreditation research. *Journal of Health Organization and Management*, 23, 255–267.

Greenhalgh, T. et al. (2009). Tensions and paradoxes in electronic patient record research: A systematic literature review using the meta-narrative method. *Milbank Quarterly*, 87(4), 729–788.

Groopman, J. E. (2007). *How Doctors Think*. Boston, MA: Houghton Mifflin.

Grote, G. (2015). Promoting safety by increasing uncertainty – Implications for risk management. *Safety Science*, 71, 71–79.

Gunasekaran, A. (1998). Agile manufacturing: enablers and an implementation framework. *International Journal of Production Research*, 36(5), 1223–1247.

Gunderson, L. H. and Holling, C. S. (eds.), (2001). *Panarchy: Understanding Transformations in Human and Natural Systems*. Washington, DC: Island Press.

Gurses, A. P. et al. (2008). Systems ambiguity and guideline compliance: A qualitative study of how intensive care units follow evidence-based guidelines to reduce healthcare associated infections. *Quality and Safety in Health Care*, 17, 351–359.

Guttmann, A. et al. (2011). Association between waiting times and short term mortality and hospital admission after departure from Emergency Department: Population based cohort study from Ontario, Canada. *BMJ*, 342, d2983.

Hale, A. R. and Borys, D. (2013a). Working to rule or working safely? Part 2: The management of safety rules and procedures. *Safety Science*, 55, 222–231.

Hale, A. R. and Borys, D. (2013b). Working to rule, or working safely? Part 1: A state of the art review. *Safety Science*, 55, 207–221.

Hale, A. R. and Swuste, P. (1998). Safety rules: Procedural freedom or action constraint? *Safety Science*, 29(3), 163–177.

Hansen, M. M. (2008). Versatile, immersive, creative and dynamic virtual 3-D healthcare learning environments: A review of the literature. *Journal of Medical Internet Research*, 10(3), e26.

Hardcastle, M. A. R. et al. (2005). An overview of structuration theory and its usefulness for nursing research. *Nursing Philosophy*, 6(4), 223–234.

Harvey, A. et al. (2012). Impact of stress on resident performance in simulated trauma scenarios. *Journal of Trauma and Acute Care Surgery*, 72(2), 497–503.

Haugaard, M. (1997). *The Constitution of Power*. Manchester, UK: Manchester University Press.

Haugaard, M. (ed.). (2002). *Power: A Reader*. Manchester, UK: Manchester University Press.

Herman Miller Healthcare and Co. (2008). *Lean Healthcare: Applying Herman Miller's Expertise to Improve Outcomes; 2008*. Online at http://www.hermanmiller.com/content/dam/hermanmiller/documents/solution_essays/SE_Lean_Healthcare.pdf, accessed 31 December 2014.

Hilligoss, B. (2014). Selling patients and other metaphors: A discourse analysis of the interpretive frames that shape Emergency Department admission handoffs. *Social Science and Medicine*, 102, 119–128.

Hilligoss, B. et al. (2015). Collaborating or selling patients? A conceptual framework of social and organizational factors in Emergency Department-to-inpatient handoff negotiations. *Joint Commission Journal on Quality and Patient Safety*, 41(3), 134–143.

Hindmarsh, D. and Lees, L. (2012). Improving the safety of patient transfer from AMU using a written checklist. *Acute Medicine*, 11, 13–17.

Holling, C. S. (2001). Understanding the complexity of economic, ecological, and social systems. *Ecosystems*, 4(5), 390–405.

Hollnagel, E. (1998). Context, cognition, and control. In Y. Waern (ed.), *Co-operation in Process Management – Cognition and Information Technology*. London, UK: Taylor & Francis (pp. 27–51).

Hollnagel, E. (2009a). *The ETTO Principle: Efficiency-Thoroughness Trade-Off. Why Things That Go Right Sometimes Go Wrong*. Farnham, Surrey, UK: Ashgate.

Hollnagel, E. (2009b). The four cornerstones of resilience engineering. In C. P. Nemeth, E. Hollnagel and S. W. A. Dekker (eds.), *Resilience Engineering Perspectives: Preparation and Restoration*. Farnham, Surrey, UK: Ashgate (pp. 117–133).

Hollnagel, E. (2010). Exploring resilience: What is it? Why is it important for healthcare? How resilience can point to critically needed solutions in healthcare. Beyond High Reliability: Improving Patient Safety Through Organizational Resilience. June 3–4, Vancouver, Canada.

Hollnagel, E. (2011a). Prologue: The scope of resilience engineering. In E. Hollnagel et al. (eds.), *Resilience Engineering in Practice: A Guidebook*. Farnham, Surrey, UK: Ashgate (pp. xxix–xxxix).

Hollnagel, E. (2011b). Epilogue: RAG – The Resilience Analysis Grid. In E. Hollnagel et al. (eds.), *Resilience Engineering in Practice: A Guidebook*. Farnham, Surrey, UK: Ashgate (pp. 275–296).

Hollnagel, E. (2011c). *How Resilient Is Your Organisation?* Online at https://hal.archives-ouvertes.fr/hal-00613986/document, accessed 10 January 2015.

Hollnagel, E. (2012a). *FRAM – The Functional Resonance Analysis Method: Modelling Complex Socio-technical Systems*. Farnham, Surrey, UK: Ashgate.

Hollnagel, E. (2012b). *A Tale of Two Safeties*. Middelfart, Denmark: University of Southern Denmark.

Hollnagel, E. (2013). Making healthcare resilient: From Safety-I to Safety-II. In E. Hollnagel, J. Braithwaite and R. L. Wears (eds.), *Resilient Health Care*. Farnham, Surrey UK: Ashgate (pp. 3–17).

Hollnagel, E. (2014a). Looking for patterns in everyday clinical work. In R. L. Wears, E. Hollnagel and J. Braithwaite (eds.), *Resilient Health Care, Volume 2: The Resilience of Everyday Clinical Work*. Farnham, Surrey, UK: Ashgate (pp. 145–161).

Hollnagel, E. (2014b). *Safety-I and Safety-II: The Past and Future of Safety Management*. Farnham, Surrey, UK: Ashgate.

Hollnagel, E. (2015a). Why is work-as-imagined different from work-as-done? In R. L. Wears, E. Hollnagel and J. Braithwaite (eds.), *Resilient Health Care, Volume 2: The Resilience of Everyday Clinical Work*. Farnham, Surrey, UK: Ashgate (pp. 249–264).

Hollnagel, E. (2015b). Looking for patterns in everyday clinical work. In R. L. Wears, E. Hollnagel and J. Braithwaite (eds.), *Resilient Health Care, Volume 2: The Resilience of Everyday Clinical Work*. Farnham, Surrey, UK: Ashgate (pp. 145–162).

Hollnagel, E. et al. (2011). *Resilience Engineering in Practice: A Guidebook*. Farnham, Surrey, UK: Ashgate.

Hollnagel, E., Braithwaite, J. and Wears, R. L. (eds.). (2013). *Resilient Health Care*. Farnham, Surrey, UK: Ashgate.

Hollnagel, E., Hounsgaard, J. and Colligan, L. (2014). *FRAM – The Functional Resonance Analysis Method: A Handbook for the Practical Use of the Method*. Middelfart, Denmark: Centre for Quality.

Hollnagel, E., Nemeth, C. P. and Dekker, S. (eds.). (2008). *Remaining Sensitive to the Possibility of Failure*. Aldershot, UK: Ashgate.

Hollnagel, E. and Woods, D. D. (2006). Epilogue: Resilience engineering precepts. In E. Hollnagel, D. D. Woods and N. G. Leveson (eds.), *Resilience Engineering: Concepts and Precepts*. Aldershot, UK: Ashgate (pp. 347–358).

Hollnagel, E., Woods, D. D. and Leveson, N. (eds.), (2006). *Resilience Engineering: Concepts and Precepts*. Aldershot, UK: Ashgate.

Homans, G. C. (1958). Group factors in worker productivity. In E. E. Maccoby, T. M. Newcomb and E. L. Hartley (eds.), *Readings in Social Psychology*. New York, NY: Holt, Rinehart and Winston (pp. 583–595).

Horne, A. and Montgomery, D. (1994). *The Lonely Leader: Monty 1944–45*. London, UK: Macmillan.

Hunte, G. S. (2010). *Creating Safety in an Emergency Department*, Vancouver, Canada: Unpublished Doctor of Philosophy Thesis, University of British Columbia. Online at https://circle.ubc.ca/handle/2429/27485, accessed 15 July 2015.

Hunte, G. S. (2015). A lesson in resilience: The 2011 Stanley Cup riot. In R. L. Wears, E. Hollnagel and J. Braithwaite (eds.), *Resilient Health Care, Volume 2: The Resilience of Everyday Clinical Work*. Farnham, Surrey, UK: Ashgate (pp. 1–9).

Hurwitz, B. and Sheikh, A. (2009). *Health Care Errors and Patient Safety*. Chichester, UK: Wiley-Blackwell/BMJ Books.

Iedema, R. et al. (2011). What prevents incident disclosure, and what can be done to promote it? *Joint Commission Journal on Quality and Patient Safety/Joint Commission Resources*, 37(9), 409–417.

Ingrassia, P. L. et al. (2012). Data collection in a live mass casualty incident simulation: Automated RFID technology versus manually recorded system. *European Journal of Emergency Medicine*, 19(1), 35–39.

Innes, G. et al. (2007). Impact of an overcapacity care protocol on Emergency Department overcrowding. *Canadian Journal of Emergency Medicine*, 9(3), 196.

Institute for Healthcare Improvement (IHI). (2014). *Implement the IHI Central Line Bundle*. Online at http://www.ihi.org/resources/Pages/Changes/Implementthe CentralLineBundle.aspx, accessed 26 December 2014.

James, W. (1890). *The Principles of Psychology*. New York, NY: H. Holt and Company.

James, B. C. and Savitz, L. A. (2011). How Intermountain trimmed health care costs through robust quality improvement efforts. *Health Affairs*, 30(6), 1185–1191.

Jangland, E. et al. (2012). The impact of an intervention to improve patient participation in a surgical care unit: A quasi-experimental study. *International Journal of Nursing Studies*, 49, 528–538.

Johnson, K. et al. (2012). Simulation to implement a novel system of care for pediatric critical airway obstruction. *Archives of Otolaryngology – Head and Neck Surgery*, 138(10), 907–911.

JCAHO. (2000). *Root cause analysis in health care: Tools and techniques*. Oakbrook Terrace: Joint Commission on Accreditation of Healthcare Organizations.

Joint Commission Centre for Transforming Healthcare. (2000). *Root Cause Analysis in Health Care: Tools and Techniques.* Oakbrook Terrace, IL: Joint Commission on Accreditation of Healthcare Organizations.

Joint Commission Centre for Transforming Healthcare. (2010). *Improving Transitions of Care: Hand-off Communications.* Oakbrook Terrace, IL: Joint Commission on Accreditation of Healthcare Organizations.

Kahneman, D. (2011). *Thinking, Fast and Slow.* London, UK: Macmillan.

Kant, I. (1781). *Critique of Pure Reason.* Trans. and ed. by Paul Guyer and Allen W. Wood. Cambridge, UK: Cambridge University Press, 1998.

Kenney, C. (2011). *Transforming Health Care: Virginia Mason Medical Centre's Pursuit of the Perfect Patient Experience.* Boca Raton, FL: CRC Press.

Kenny, K. E. (2014). Blaming deadmen: Causes, culprits, and chaos in accounting for technological accidents. *Science, Technology and Human Values,* 40(4), 539–563.

King, D. L., Ben-Tovim, D. I. and Bassham, J. (2006). Redesigning Emergency Department patient flows: Application of Lean thinking to health care. *Emergency Medicine Australasia,* 18(4), 391–397.

Kirkpatrick, D. L. (1978). Evaluating in-house training programs. *Training and Development Journal,* 32(9), 6–9.

Kitson, A. et al. (2013). What are the core elements of patient-centred care? A narrative review and synthesis of the literature from health policy, medicine and nursing. *Journal of Advanced Nursing,* 69, 4–15.

Klass, P. (2007). *Treatment Kind and Fair: Letters to a Young Doctor.* New York, NY: Basic Books.

Kobayashi, L. et al. (2006). Portable advanced medical simulation for new Emergency Department testing and orientation. *Academic Emergency Medicine,* 13(6), 691–695.

Kobayashi, L. et al. (2013). Use of in situ simulation and human factors engineering to assess and improve Emergency Department clinical systems for timely telemetry-based detection of life-threatening arrhythmias. *BMJ Quality and Safety,* 22(1), 72–83.

Korzybski, A. (1931). A non-Aristotelian system and its necessity for rigour in mathematics and physics. Paper presented before the American Mathematical Society at the New Orleans, Louisiana, meeting of the American Association for the Advancement of Science, 28 December 1931. Reprinted in *Science and Sanity,* 1933, pp. 747–761.

Kuhn, T. (1962). *The Structure of Scientific Revolutions.* Chicago, IL: University of Chicago press Ltd.

Kushniruk, A. W. and Borycki, E. M. (2014). Designing and conducting low-cost in-situ clinical simulations: A methodological approach. *Studies in Health Technology and Informatics,* 205, 890–894.

Landman, A. et al. (2014). Efficiency and usability of a near field communication-enabled tablet for medication administration. *Journal of Medical Internet Research mHealth and uHealth,* 2(2), e26.

Lanir, Z. (1983). *Fundamental Surprise – Israeli Lessons.* Online at http://www.articlesbase.com/economics-articles/fundamental-surprise-israeli-lessons-2784076.html, accessed 20 February 2015.

Laskowski, M. et al. (2009). Models of Emergency Departments for reducing patient waiting times. *PLoS One,* 4(7), e6127.

Laugaland, K. and Aase, K. (2014). The demands imposed by a health care reform on clinical work in transitional care of the elderly: A multi-faceted Janus. In R. L. Wears, E. Hollnagel and J. Braithwaite (eds.), *Resilient Health Care, Volume 2: Resilience of Everyday Clinical Work*. Farnham, Surrey, UK: Ashgate (pp. 39–57).

Laursen, M. L., Gersten, F. and Johansen, J. (2003). *Applying Lean Thinking in Hospitals – Exploring Implementation Difficulties*. Aalborg University, Center for Industrial Production. Online at http://www.lindgaardconsulting.dk, assessed 16 July 2015.

Lawton, R. and Parker, D. (2002). Barriers to incident reporting in a healthcare system. *Quality and Safety in Health Care*, 11, 15–18.

Lazare, A. (2004). *On Apology*. Oxford, UK: Oxford University Press.

Lazarus, R. S. and Folkman, S. (1984). *Stress, Appraisal and Coping*. New York, NY: Springer.

Lebel, L. et al. (2006). Governance and the capacity to manage resilience in regional social-ecological systems. *Ecology and Society*, 11(1), 19.

LeBlanc, V. et al. (2008). Examination stress leads to improvements on fundamental technical skills for surgery. *American Journal of Surgery*, 196(1), 114–119.

LeBlanc, V. R. and Bandiera, G. W. (2007). The effects of examination stress on the performance of emergency medicine residents. *Medical Education*, 41(6), 556–564.

Leplat, J. (1975). La charge de travail dans la régulation de l'activité: Quelques application sur les opérateurs vieillissants. In A. Laville, C. Teiger and A. Wisner (eds.), *Age et Contraintes de Travail*. Jouy en Josas, France: Editions Scientifiques (pp. 209–224).

Leplat, J. (1990). Relations between task and activity: Elements for elaborating a framework for error analysis. *Ergonomics*, 33(10–11), 1389–1402.

Lichtenstein, B. B. et al. (2006). *Complexity Leadership Theory: An Interactive Perspective on Leading in Complex Adaptive Systems*. Lincoln, NE: University of Nebraska.

Liker, J. (2004). *The Toyota Way: 14 Management Principles from the World's Greatest Manufacturer*. New York, NY: McGraw-Hill.

Lindblom, C. E. (1959). The science of "muddling through." *Public Administration Review*, 19, 79–88.

Lindblom, C. E. (1979). Still muddling, not yet through. *Public Administration Review*, 39(6), 517–526.

Locock, L. (2003). Healthcare redesign: Meaning, origins and application. *Quality and Safety in Health Care*, 12, 53–57.

Lodge, A. and Bamford, D. (2008). New development: Using Lean techniques to reduce radiology waiting times. *Public Money and Management*, 28(1), 49–52.

Longtin, Y. et al. (2010). Patient participation: Current knowledge and applicability to patient safety. *Mayo Clinic Proceedings*, 85, 53–62.

Lundberg, J., Rollenhagen, C. and Hollnagel, E. (2009). What-You-Look-for-Is-What-You-Find – The consequences of underlying accident models in eight accident investigation manuals. *Safety Science*, 47(10), 1297–1311.

MacDonald, N. and Attaran, A. (2009). Medical errors, apologies and apology laws. *Canadian Medical Association Journal*, 180(1), 11–12.

Machiavelli, N. (1975). *The Prince*. Translated with an Introduction by George Bull. London, UK: Penguin Books.

Madhok, R. et al. (2014). How to protect the 'second victim' of adverse events. *Health Service Journal*, 5, 1–4.

Malec, J. F. et al. (2007). The Mayo High Performance Teamwork Scale: Reliability and validity for evaluating key Crew Resource Management skills. *Simulation in Healthcare*, 2(4), 4–10.

March, J. G. (1991). Exploration and exploitation in organizational learning. *Organization Science*, 2(1), 71–87.

March, J. G., Sproull, L. S. and Tamuz, M. (1991). Learning from samples of one or fewer. *Organization Science*, 2(1), 1–13.

Martin, K. and Osterling, M. (2014). *Value Stream Mapping: How to Visualize Work and Align Leadership for Organizational Transformation.* New York, NY: McGraw-Hill.

May, C. R. et al. (2014). Rethinking the patient: Using Burden of Treatment Theory to understand the changing dynamics of illness. *BMC Health Services Research*, 14(1), 1–11.

Mayo, E. (1931). *Social Problems of an Industrial Civilization.* Boston, MA: Division of Research, Graduate School of Business Administration, Harvard University.

Mazzocato, P. et al. (2010). Lean Thinking in healthcare: A realist review of the literature; *Quality Safety in Health Care*, 19, 376–382.

McKee, P. and Macleod, O. (2012). *Root Cause Analysis of MRSA/MSSA Bacteraemias and Clostridium Difficile Infections (CDI) – Protocol.* Northern Health and Social Care Trust. Online at http://www.northerntrust.hscni.net/pdf/Root_Cause_Analysis_of_MRSA_MSSA_Bacteraemias_and_Clostridium_Difficile_Infections.pdf, accessed 23 January 2015.

Meeks, D. W. et al. (2014). Exploring the sociotechnical intersection of patient safety and electronic health record implementation. *Journal of American Medical Informatics Association*, 21, e28–e34.

Merton, R. (1936). The unanticipated consequences of social action. *American Sociological Review*, 1, 894–904.

Michallet, B. (2009). Resilience: Perspectives historiques, defis théoriques et enjeux cliniques. *Frontières*, 22(1–2), 10–18.

Mikulincer, M. and Solomon, Z. (1989). Causal attribution, coping strategies and combat post-traumatic stress disorders. *European Journal of Personality*, 3(1), 269–284.

Miller, G. A., Galanter, E. and Pribram, K. H. (1960). *Plans and the Structure of Behavior.* New York, NY: Holt, Rinehart & Winston.

Miller, K. K. et al. (2008). In situ simulation: A method of experiential learning to promote safety and team behavior. *Journal of Perinatal and Neonatal Nursing*, 22(2), 105–113.

Mumber, M. (2014). Effect of mindfulness training on mindfulness level in the workplace and patient safety culture as a part of error prevention in radiation oncology practice: A pilot study. *International Journal of Radiation Oncology, Biology, Physics*, 90, S748.

Mumford, L. (1934). *Technics and Civilization.* New York, NY: Harcourt, Brace and Company.

Nakajima, K. (2015). Blood transfusion with health information technology in emergency settings from a Safety-II perspective. In R. L. Wears, E. Hollnagel and J. Braithwaite (eds.), *Resilient Health Care, Volume 2: The Resilience of Everyday Clinical Work.* Farnham, Surrey, UK: Ashgate (pp. 99–113).

National Health Service (UK). (2013). *The NHS Constitution.* Online at http://www.nhs.uk/choiceintheNHS/Rightsandpledges/NHSConstitution/Documents/2013/the-nhs-constitution-for-england-2013.pdf, accessed 24 December 2014.

Nelson, D. R., Adger, W. N. and Brown, K. (2007). Adaptation to environmental change: Contributions of a resilience framework. *Annual Review of Environment and Resources*, 32(1), 395–419.

Nemeth, C. P. et al. (2008). Minding the gaps: Creating resilience in health care. In K. Henriksen et al. (eds.), *Advances in Patient Safety: New Directions and Alternative Approaches*. Rockville, MD: Agency for Healthcare Research and Quality.

Nemeth, C. P., Cook, R. I. and Wears, R. L. (2007). Studying the technical work of emergency care. *Annals of Emergency Medicine*, 50(4), 384–386.

Nugus, P., Bridges, J. and Braithwaite, J. (2009). Selling patients. *BMJ*, 339:b5201.

Nugus, P. et al. (2010). Integrated care in the Emergency Department: A complex adaptive systems perspective. *Social Science and Medicine*, 71(11), 1997–2004.

Nugus, P. et al. (2011). Work pressure and patient flow management in the Emergency Department: Findings from an ethnographic study. *Academic Emergency Medicine*, 18(10), 1045–1052.

Nugus, P. et al. (2014). The Emergency Department "carousel": An ethnographically-derived model of the dynamics of patient flow. *International Emergency Nursing*, 22, 3–9.

Nugus, P., Sheikh, M. and Braithwaite, J. (2012). Structuring emergency care: Policy and organisational behavioural dimensions. In H. Dickinson and R. Mannion (eds.), *The Reform of Health Care: Shaping, Adapting and Resisting Policy Developments*. London, UK: Palgrave Macmillan (pp. 151–163).

Nyssen, A. S. (2011). From myopic coordination to resilience in socio-technical systems. A case study in a hospital. In E. Hollnagel et al. (eds.), *Resilience Engineering in Practice: A Guidebook*. Farnham, Surrey, UK: Ashgate (pp. 219–236).

Nyssen, A. S. and Blavier, A. (2013). Investigating expertise, flexibility and resilience in socio-technical environments: A case study in robotic surgery. In E. Hollnagel, J. Braithwaite and R. L. Wears (eds.), *Resilient Health Care*. Farnham, Surrey, UK: Ashgate (pp. 97–110).

Nyssen, A. S. et al. (2004). Reporting systems in health care from a case-by-case experience to a general framework: An example in anesthesia. *European Journal of Anaesthesiology*, 21(10), 757–765.

Nyssen, A. S. et al. (2012). Systèmes de retour d'expérience: Faut-il vraiment copier l'industrie? *Risques and Qualité*, 9(2), 85–91.

O'Connor, P. et al. (2008). Crew Resource Management training effectiveness: A meta-analysis and some critical needs. *International Journal of Aviation Psychology*, 18, 353–368.

Ombredane, A. and Faverge, J. M. (1955). *L'analyse du Travail*. Paris, France: Presses Universitaires de France.

Ostrom, E. (1990). *Governing the Commons: The Evolution of Institutions for Collective Action*. Cambridge, UK: Cambridge University Press.

Ostrom, E. (1999a). Polycentricity, complexity, and the commons. *The Good Society*, 9(2), 37–41.

Ostrom, E. (1999b). Coping with tragedies of the commons. *Annual Review of Political Science*, 2, 493–535.

Ostrom, E. (2007). A diagnostic approach for going beyond panaceas. *Proceedings of the National Academy of Sciences*, 104(39), 15181–15187.

Packwood, T., Pollitt, C. and Roberts, S. (1998). Good medicine? A case study of business process re-engineering in a hospital. *Policy and Politics*, 26(4), 401–415.

Parunak, H. V. D. (2006). A survey of environments and mechanisms for human-human stigmergy. In D. Weyns, H. Van Dyke Parunak and F. Michel (eds.), *Environments for Multi-Agent Systems II*. Berlin, Germany: Springer (pp. 163–186).

Pasquini, A. et al. (2011). Requisites for successful incident reporting in resilient organisations. In E. Hollnagel et al. (eds.), *Resilience Engineering in Practice: A Guidebook*. Farnham, Surrey, UK: Ashgate (pp. 237–256).

Patterson, E. S. et al. (2006a). Gaps and resilience. In M. S. Bogner (ed.), *Human Error in Medicine* (2nd ed.). Hillsdale, NJ: Lawrence Erlbaum.

Patterson, E. S. et al. (2006b). Three key levers for achieving resilience in medication delivery with information technology. *Journal of Patient Safety*, 2(1), 33.

Patterson, M. D. et al. (2013). In situ simulation: Detection of safety threats and team-work training in a high risk Emergency Department. *BMJ Quality and Safety*, 22(6), 468–477.

Patterson, M. D., Blike, G. T. and Nadkarni, V. M. (2008). In situ simulation: Challenges and results. In K. Henriksen et al. (eds.), *Advances in Patient Safety: New Directions and Alternative Approaches (Volume 3: Performance and Tools)*, Publication 08-0034-3. Rockville, MD: Agency for Healthcare Research and Quality.

Perrow, C., Wilensky, H. L. and Reiss, A. J. (1972). *Complex Organizations: A Critical Essay*. Glenview, IL: Scott, Foresman.

Piaget, J. (1967). *Biologie et Connaissance*. Paris, France: Gallimard.

Pilgrim, D., Tomasini, F. and Vassilev, I. (2011). *Examining Trust in Healthcare: A Multidisciplinary Perspective*. Houndmills, Basingstoke, Hampshire, UK: Palgrave Macmillan.

Plato (514a–540a). *The Republic*. In Jowett, B. (ed.) (1941). *Plato's The Republic*. New York, NY: The Modern Library.

Powell, W. W., Koput, K. W. and Smith-Doerr, L. (1996). Interorganisational collaboration and the locus of innovation: Networks of learning in biotechnology. *Administrative Science Quarterly*, 41, 116–145.

Prince, C. and Salas, E. (1993). Training and research for teamwork in the military aircrew. In E. Wiener et al. (eds.), *Cockpit Resource Management*. San Diego, CA: Academic Press (pp. 337–366).

Proudlove, N., Moxham, C. and Boaden, R. (2008). Lessons for Lean in healthcare from using Six Sigma in the NHS. *Public Money and Management*, 28(1), 27–43.

Radnor, Z. J., Holweg, M. and Waring, J. (2012). Lean in healthcare: The unfilled promise? *Social Science and Medicine*, 74, 364–371.

Radnor, Z. J. and Walley, P. (2008). Learning to walk before we try to run: Adapting Lean for the public sector. *Public Money and Management*, 28(1), 13–20.

Rasmussen, J. and Lind, M. (1981). *Coping with Complexity*. Roskilde, Denmark: Risø National Laboratory.

Reason, J. (1997). *Managing the Risks of Organizational Accidents*. Farnham, Surrey, UK: Ashgate.

Reason, J. (2000). Human error: Models and management. *BMJ*, 320, 768–770.

Reznek, M. et al. (2003). Emergency medicine crisis resource management (EMCRM): Pilot study of a simulation-based crisis management course for emergency medicine. *Academic Emergency Medicine*, 10, 386–389.

Richards, T. et al. (2013). Let the patient revolution begin. *BMJ*, 346, f2614.

Righi, A. W. and Saurin, T. A. (2015). Complex socio-technical systems: Characterization and management guidelines. *Applied Ergonomics*, 50, 19–30.

Riley, W. et al. (2010). Detecting breaches in defensive barriers using in situ simulation for obstetric emergencies. *Quality and Safety in Health Care*, 19(suppl. 3), 53–56.

Roberts, K. H. et al. (2005). A case of the birth and death of a high reliability healthcare organisation. *Quality and Safety in Health Care*, 14(3), 216–220.

Robson, R. (2013). Resilient health care. In E. Hollnagel, J. Braithwaite and R. L. Wears (eds.), *Resilient Health Care*. Farnham, Surrey, UK: Ashgate (pp. 191–203).

Robson, R. (2015). ECW in complex adaptive systems. In R. L. Wears, E. Hollnagel and J. Braithwaite (eds.), *Resilient Health Care, Volume 2: The Resilience of Everyday Clinical Work*. Farnham, Surrey, UK: Ashgate (pp. 177–188).

Rogers, E. M. and Shoemaker, F. F. (1971). *Communication of Innovations: A Cross-Cultural Approach*. New York, NY: Free Press.

Rother, M. and Shook, J. (1998). *Learning to See: Value Stream Mapping to Add Value and Eliminate Muda*. Cambridge, MA: Lean Enterprise Institute.

Rust, T. C. (2013). *Dynamic Analysis of Healthcare Service Delivery: Application of Lean and Agile Concepts*. Unpublished Doctor of Philosophy Thesis, Worcester Polytechnic Institute, Worcester, MA. Online at https://www.wpi.edu/Pubs/ETD/Available/etd-043013-140924/unrestricted/Rust-Dissertation-May-3-2013.pdf, accessed 15 July 2015.

Salas, E. et al. (2008a). Does team training improve team performance? A meta-analysis. *Human Factors*, 50, 903–933.

Salas, E. et al. (2008b). Does team training work: Principles for health care. *Academic Emergency Medicine*, 15, 1002–1009.

Sari, A. B. et al. (2007). Extent, nature and consequences of adverse events: Results of a retrospective case note review in a large NHS hospital. *Quality and Safety in Health Care*, 16, 434–439.

Saurin, T. A., Rooke, J. and Koskela, L. (2013). A complex systems theory perspective of Lean production. *International Journal of Production Research*, 51, 5824–5838.

Schubert, C. C. et al. (2015). Patients as a source of resilience. In R. L. Wears, E. Hollnagel and J. Braithwaite (eds.), *Resilient Health Care, Volume 2: The Resilience of Everyday Clinical Work*. Farnham, Surrey, UK: Ashgate (pp. 207–223).

Schulman, P. R. (1993). The negotiated order of organizational reliability. *Administration and Society*, 25(3), 353–372.

Schwabe, L., Wolf, O. T. and Oitzl, M. S. (2010). Memory formation under stress: Quantity and quality. *Neuroscience and Biobehavioral Reviews*, 34(4), 584–591.

Scott, J. C. (2012). *Two Cheers for Anarchism*. Princeton, NJ: Princeton University Press.

Seddon, J. and Brand, C. (2008). Debate: Systems thinking and public sector performance. *Public Money and Management*, 28(1), 7–9.

Sexton, J. B. et al. (2006). The Safety Attitudes Questionnaire: Psychometric properties, benchmarking data, and emerging research. *BMC Health Services Research*, 6, 44.

Shah, R. and Ward, P. (2007). Defining and developing measures of Lean production. *Journal of Operations Management*, 25, 785–805.

Sharpe, V. A. and Faden, A. I. (1998). *Medical Harm: Historical, Conceptual, and Ethical Dimensions of Iatrogenic Illness*. Cambridge, UK: Cambridge University Press.

Sheps, S. B., Cardiff, K. and Robson, R. (2011). Patient safety: A wake up call. In *Proceedings of the Fourth Resilience Engineering Symposium*, Sophia Antipolis, France, 8–10 June (pp. 248–255).

Shiffman, R. N. (1997). Representation of clinical practice guidelines in conventional and augmented decision tables. *Journal of the American Medical Informatics Association*, 4(5), 382–393.

Shimai, Y., Nagahama, M. and Nakajima, K. (2011). An analysis of patient identification procedures in intravenous medication administration using health information technology [in Japanese]. *Health Information Management*, 23(2), 276.

Shojania, K. G. (2008). The frustrating case of incident-reporting systems. *Quality and Safety in Health Care*, 17, 400–402.

Sibinga, E. M. and Wu, A. W. (2010). Clinician mindfulness and patient safety. *Journal of the American Medical Association*, 304, 2532–2533.

Simon, H. A. (1956). Rational choice and the structure of the environment. *Psychological Review*, 63(2), 129–138.

Smalley, A. (2004). *Creating Level Pull*. Cambridge, MA: Lean Enterprise Institute.

Smith, M. et al. (2013). Resilient actions in the diagnostic process and system performance. *BMJ Quality and Safety*, 22(12), 1005–1012.

Snook, S. A. (2000). *Friendly Fire: The Accidental Shoot-Down of US Black Hawks over Northern Iraq*. Princeton, NJ: Princeton University Press.

Spear, S. (2005). Fixing healthcare from the inside, today. *Harvard Business Review*. September. Online at https://hbr.org/2005/09/fixing-health-care-from-the-inside-today, accessed 15 July 2015.

Spencer, M., Dineen, R. and Phillip, A. (2013). *Co-producing Services – Co-creating Health*. 1000 Lives Plus Programme. Online at http://www.1000livesplus.wales.nhs.uk/sitesplus/documents/1011/T4I%20%288%29%20Co-production.pdf, accessed 15 July 2015.

Stein, J. G. (2002). *The Cult of Efficiency*. Toronto, Canada: House of Anansi Press and Groundwood Books.

Stephens, R., Woods, D. D. and Patterson, E. (2015). Patient boarding in the Emergency Department as a symptom of complexity-induced risks. In R. L. Wears, E. Hollnagel and J. Braithwaite (eds.), *Resilient Health Care, Volume 2: The Resilience of Everyday Clinical Work*. Farnham, Surrey, UK: Ashgate (pp. 129–143).

Storr, J., Wigglesworth, N. and Kilpatrick, C. (2013). *Integrating Human Factors with Infection Prevention and Control*. The Health Foundation. Online at http://www.health.org.uk/sites/default/files/IntegratingHumanFactorsWithInfectionAndPreventionControl.pdf, accessed 23 January 2015.

Strauss, A. et al. (1963). The hospital and its negotiated order. In E. Friedson (ed.), *The Hospital in Modern Society*. New York, NY: Free Press of Glencoe.

Strauss, A. L. et al. (1997). *Social Organization of Medical Work* (2nd ed.). Chicago, IL: University of Chicago Press.

Sujan, M. A. (2012). A novel tool for organisational learning and its impact on safety culture in a hospital dispensary. *Reliability Engineering and System Safety*, 101, 21–34.

Sujan, M. A. et al. (2011a). Hassle in the dispensary: Pilot study of a proactive risk monitoring tool for organisational learning based on narratives and staff perceptions. *BMJ Quality and Safety*, 20, 549–556.

Sujan, M. A. et al. (2011b). Resilience as individual adaptation: Preliminary analysis of a hospital dispensary. Fourth Workshop HCP Human Centered Processes, Genoa, Italy, 10–11 February.

Sujan, M. A., Spurgeon, P. and Cooke, M. (2015a). Translating tensions into safe practices through dynamic trade-offs: The secret second handover. In R. L. Wears, E. Hollnagel and J. Braithwaite (eds.), *Resilient Health Care, Volume 2: Resilience of Everyday Clinical Work*. Farnham, Surrey, UK: Ashgate (pp. 11–22).

Sujan, M. A., Spurgeon, P. and Cooke, M. (2015b). Managing competing organizational priorities in clinical handover across organizational boundaries. *Journal of Health Services Research and Policy*, 20, 17–25.

Sujan, M., Spurgeon, P. and Cooke, M. (2015c). The role of dynamic trade-offs in creating safety – A qualitative study of handover across care boundaries in emergency care. *Reliability Engineering and System Safety*, 14, 54–62.

Susi, T. and Ziemke, T. (2001). Social cognition, artefacts, and stigmergy: A comparative analysis of theoretical frameworks for the understanding of artefact-mediated collaborative activity. *Cognitive Systems Research*, 2(4), 273–290.

Sutcliffe, K. and Weick, K. (2013). Mindful organising and resilient health care. In E. Hollnagel, J. Braithwaite and R. L. Wears (eds.), *Resilient Health Care*. Farnham, Surrey, UK: Ashgate (pp. 145–156).

Tamuz, M., Franchois, K. E. and Thomas, E. J. (2011). What's past is prologue: Organizational learning from a serious patient injury. *Safety Science*, 49, 75–82.

Tariman, J. et al. (2010). Preferred and actual participation roles during health care decision making in persons with cancer: A systematic review. *Annals of Oncology*, 21, 1145–1151.

Thompson, J. E. et al. (2011). Using the ISBAR handover tool in junior medical officer handover: A study in an Australian tertiary hospital. *Postgraduate Medical Journal*, 87, 340–344.

Timmermans, S. (2010). Evidence-based medicine: Sociological explorations. In C. E. Bird, P. Conrad and A. M. Fremont (eds.), *Handbook of Medical Sociology*. Nashville, TN: Vanderbilt University Press (pp. 309–323).

Townsend, A. S. (2013). *Safety Can't Be Measured: An Evidence-Based Approach to Risk Reduction*. Farnham, Surrey, UK: Gower.

Tragardh, B. and Lindberg, K. (2004). Curing a meagre health care system by Lean methods – Translating chains of care in the Swedish health care sector. *International Journal of Health Planning and Management*, 19(4), 383–398.

Tuchman, B. W. (1976). *The Guns of August*. New York, NY: Bantam Books.

Uhl-Bien, M., Marion, R. and McKelvey, B. (2007). Complexity leadership theory: Shifting leadership from the industrial age to the knowledge era. *Leadership Quarterly*, 18, 298–318.

Vest, J. R. and Gamm L. D. (2009). A critical review of the research literature on Six Sigma, Lean and StuderGroup's Hardwiring Excellence in the United States: The need to demonstrate and communicate the effectiveness of transformation strategies in healthcare. *Implementation Science*, 35(4), 19.

Vicente, K. J. (1999). *Cognitive Work Analysis: Towards Safe, Productive, and Healthy Computer-Based Work*. Mahwah, NJ: Lawrence Erlbaum.

Vincent, C., Neale, G. and Woloshynowych, M. (2001). Adverse events in British hospitals: Preliminary retrospective record review. *BMJ*, 322, 517–519.

Von Buren, H. (2013). *Análise de Acidentes do Trabalho Utilizando Modelos Criados para Sistemas Sócio-técnicos Complexos: Estudo de caso Empregando o Método FRAM*. Unpublished MSc Thesis, Federal University of Rio de Janeiro, Rio de Janeiro, Brazil.

Voß, J. P., Bauknecht, D. and Kemp, R. (2006). *Reflexive Governance for Sustainable Development*. Gloucestershire, UK: Edward Elgar.

Vries, J. and Huijsman, R. (2011). Supply chain management in health services: an overview. *Supply Chain Management: An International Journal*, 16(3), 159–165.

Walker, M. U. (2006). *Moral Repair: Reconstructing Moral Relations after Wrongdoing*. Cambridge, UK: Cambridge University Press.

Wallace, B. and Ross, A. J. (2006) *Beyond Human Error: Taxonomies and Safety Science*. Boca Raton, FL: CRC Press.

Waring, J. J. and Bishop, S. (2010). Lean healthcare: Rhetoric, ritual and resistance. *Social Science and Medicine*, 71(7), 1332–1340.

Warr, P. and Bunce, D. (1995). Trainee characteristics and the outcomes of open learning. *Personnel Psychology*, 48, 347–375.

Wears, R. L. (2010). Exploring the dynamics of resilience. In *Proceedings of Human Factors and Ergonomics Society 54th Annual Meeting*, San Francisco, CA, 27 September–1 October, Human Factors and Ergonomics Society (pp. 394–398).

Wears, R. L. (2015). Standardisation and its discontents. *Cognition, Technology and Work*, 17(1), 89–94.

Wears, R. L., Hollnagel, E. and Braithwaite, J. (2015). *Resilient Health Care, Volume 2: The Resilience of Everyday Clinical Work*. Farnham, Surrey, UK: Ashgate.

Wears, R. L. and Hunte, G. S. (2014). Seeing patient safety 'Like a State'. *Safety Science*, 64, 50–57.

Wears, R. L., Perry, S. and McFauls, A. (2006). Free fall – A case study of resilience, its degradation, and recovery, in an Emergency Department. Paper presented at the Second Resilience Engineering Symposium, Juan-les-Pins, France, 8–10 November (pp. 325–332). Online at http://www.resilience-engineering-association.org/download/resources/symposium/symposium-2006(2)/Wears_et_al.pdf, accessed 15 July 2015.

Wears, R. and Vincent, C. (2013). Relying on resilience: Too much of a good thing? In E. Hollnagel, J. Braithwaite and R. L. Wears, (eds.), *Resilient Health Care*. Farnham, Surrey, UK: Ashgate (pp. 153–163).

Weaver, S. J. et al. (2010). The anatomy of health care team training and the state of practice: A critical review. *Academic Medicine*, 85, 1746–1760.

Weaver, S. J., Dy, S. M. and Rosen, M. A. (2014). Team-training in healthcare: A narrative synthesis of the literature. *BMJ Quality and Safety*, 23, 359–372.

Weber, M. (1922/1978). *Economy and Society: An Outline of Interpretive Sociology*. G. Roth and C. Wittich (eds.). Berkley, CA: University of California Press.

Weick, K. and Sutcliffe, K. (2007). *Managing the Unexpected: Resilient Performance in an Age of Uncertainty*. San Francisco, CA: Jossey-Bass.

Weingart, S. N. et al. (2011). Hospitalized patients' participation and its impact on quality of care and patient safety. *International Journal for Quality in Health Care*, 23, 269–277.

Westbrook, J. I. et al. (2015). What are incident reports telling us? A comparative study at two Australian hospitals of medication errors identified at audit, detected by staff and reported to an incident system. *International Journal for Quality in Health Care*, 27, 1–9.

Westrum, R. (1993). Cultures with requisite imagination. In J. A. Wise, V. D. Hopkin and P. Stager (eds.), *Verification and Validation of Complex Systems: Human Factors Issues*. Berlin, Germany: Springer Verlag (pp. 401–416).

Westrum, R. (2006). A typology of resilience situations. In E. Hollnagel, D. D. Woods and N. G. Leveson (eds.), *Resilience Engineering: Concepts and Precepts*. Farnham, Surrey, UK: Ashgate (pp. 55–65).

Wheeler, D. S. et al. (2013). High-reliability emergency response teams in the hospital: Improving quality and safety using in situ simulation training. *BMJ Quality and Safety*, 22(6), 507–514.

Whetten, D. A. (1995). Introduction to the Series. In K. E. Weick (ed.), *Sensemaking in Organizations*. Thousand Oaks, CA: Sage (pp. vii–ix).

Wiley, J. et al. (2014). Shared decision-making: The perspectives of young adults with type 1 diabetes mellitus. *Patient Preference and Adherence*, 8, 423–435.

Wilson, R. (2001). *The Politics of Truth and Reconciliation in South Africa: Legitimizing the Post-Apartheid*. New York, NY: Cambridge University Press.

Wittgenstein, L. (2009). *Philosophical Investigations* (Rev. 4th ed. by P. M. S. Hacker and J. Schulte). New York, NY: Macmillan.

Woods, D. D. (2014). Innovation: The flip side of resilience. In J. Bloomberg (ed.), *Forbes*. 9 November. Online at http://www.forbes.com/sites/jasonbloomberg/2014/09/23/innovation-the-flip-side-of-resilience/, accessed 16 July 2015.

Woods, D. D. (2015). Four concepts for resilience and the implications for the future of resilience engineering. *Reliability Engineering and Systems Safety*, 141, 5–9.

Woods, D. D. and Cook, R. I. (2002). Nine steps to move forward from error. *Cognition, Technology and Work*, 4, 137–144.

Woods, D. D. and Hollnagel, E. (2006). Prologue: Resilience engineering concepts. In E. Hollnagel, D. D. Woods and N. G. Leveson (eds.), *Resilience Engineering: Concepts and Precepts*. Aldershot, UK: Ashgate (pp. 1–16).

Woods, D. D. and Wreathall, J. (2008). Stress–strain plots as a basis for assessing system resilience. In E. Hollnagel, C. P. Nemeth and S. W. A. Dekker (eds.), *Resilience Engineering Perspective: Remaining Sensitive to the Possibility of Failure*. Aldershot, UK: Ashgate (pp. 143–158).

World Health Organization. (2008). *World Alliance for Patient Safety Progress Report 2006–2007*. Online at http://www.who.int/patientsafety/information_centre/documents/en/, accessed 24 December 2014.

Young, T. P. and McClean, S. I. (2008). A critical look at Lean Thinking in healthcare. *Quality and Safety in Health Care*, 17(5), 382–386.

Zorbas, E. (2004). Reconciliation in post-genocide Rwanda. *African Journal of Legal Studies*, 1(1), 29–52.

Index